SpringerBriefs in Physics

SpringerBriefs in Physics are a series of slim high-quality publications encompassing the entire spectrum of physics. Manuscripts for SpringerBriefs in Physics will be evaluated by Springer and by members of the Editorial Board. Proposals and other communication should be sent to your Publishing Editors at Springer.

Featuring compact volumes of 50 to 125 pages (approximately 20,000–45,000 words), Briefs are shorter than a conventional book but longer than a journal article. Thus, Briefs serve as timely, concise tools for students, researchers, and professionals.

Typical texts for publication might include:

- A snapshot review of the current state of a hot or emerging field
- A concise introduction to core concepts that students must understand in order to make independent contributions
- An extended research report giving more details and discussion than is possible in a conventional journal article
- A manual describing underlying principles and best practices for an experimental technique
- An essay exploring new ideas within physics, related philosophical issues, or broader topics such as science and society

Briefs allow authors to present their ideas and readers to absorb them with minimal time investment.

Briefs will be published as part of Springer's eBook collection, with millions of users worldwide. In addition, they will be available, just like other books, for individual print and electronic purchase.

Briefs are characterized by fast, global electronic dissemination, straightforward publishing agreements, easy-to-use manuscript preparation and formatting guidelines, and expedited production schedules. We aim for publication 8–12 weeks after acceptance.

More information about this series at http://www.springer.com/series/8902

Neophytos Neophytou

Theory and Simulation Methods for Electronic and Phononic Transport in Thermoelectric Materials

 Springer

Neophytos Neophytou
School of Engineering
University of Warwick
Coventry, UK

ISSN 2191-5423   ISSN 2191-5431   (electronic)
SpringerBriefs in Physics
ISBN 978-3-030-38680-1   ISBN 978-3-030-38681-8   (eBook)
https://doi.org/10.1007/978-3-030-38681-8

This Springer imprint is published by the registered company Springer Nature Switzerland AG
The registered company address is: Gewerbestrasse 11, 6330 Cham, Switzerland

To Stella, Markos and Lukas

Preface

Up until the late 1990s only a handful of materials, the most common of which were Bi_2Te_3, PbTe, SiGe, and their alloys, exhibited a thermoelectric figure of merit $ZT \sim 1$, corresponding to 10% of the Carnot efficiency. The field of thermoelectric materials for energy conversion and generation, however, has undergone an enormous progress over the last three decades. This is owed to the experimental realization of numerous complex bandstructure materials, their alloys, and their nanostructures. These developments allowed for large performance improvements across materials and temperatures, and constitute the most significant advancements in the field in the last 50 years. The number of literature publications in the field has drastically increased over the last years, and the ZT figures-of-merit have consistently reached values $ZT > 2$, even up to 2.6 in special cases. The complexity in understanding and favorably designing the material properties for even further improvements calls for strong support from advanced theory and simulation in the areas of electronic and thermal transport. There are numerous techniques routinely employed in other science and engineering fields, namely, the electronic device areas, which can be adopted to the needs of the thermoelectric community.

The purpose of this brief is to describe such important and useful electronic and phononic transport techniques, for the more accurate simulation of thermoelectric properties for both: (i) complex bandstructure materials and (ii) nanostructured materials. The techniques needed have to be able to describe electronic transport taking into account the full energy dependencies of the bandstructure and the scattering physics. They need to merge length scales and physical phenomena, i.e., from atomistic to continuum and from quantum mechanical to fully diffusive, regimes that can co-exist in a typical nanostructured thermoelectric material. They also need to describe transport in arbitrary disordered geometries. Such techniques are typically found and utilized by transport communities and device communities, which are somewhat disconnected from the materials science and the thermoelectric communities. Thus, the brief is an attempt to provide a summary of some of the methods I find particularly useful for the interested researcher. As the theory of each method is well established in the literature through multiple sources that include papers, books, and lectures that one can easily find online (for example, on nanoHUB.org), in this brief I do not provide a detailed theory and derivations of all methodologies. Instead, I focus on numerical implementations and only provide the basic theory needed to understand those implementations. The brief introduces the theory, computational complexities, the appropriateness of each method for different material cases, its strengths, capabilities, and weaknesses. The brief should be perceived as a guide to computational researchers and graduate students engaged in the field of computation of electro-thermal transport in thermoelectric materials, for whom I hope it will prove to be a valuable resource. It is presented in a way to be accessible to junior computational scientists, however, includes enough detail and goes into the necessary depth that is needed for more involved, advanced simulations.

In writing this brief, I particularly would like to thank my co-workers at the University of Warwick Dhritiman Chakraborty, Vassilios Vargiamidis, Laura de Sousa Oliveira, Chathurangi Kumarasinghe, Samuel Foster, and Patrizio Graziosi for helping me in the preparation of the material. I also would like to acknowledge numerous helpful discussions with Mischa Thesberg and Hans Kosina from the Vienna University of Technology.

Coventry, UK   Neophytos Neophytou

Acknowledgements Neophytos Neophytou acknowledges funding from the European Research Council (ERC) under the European Union's Horizon 2020 Research and Innovation Programme (Grant Agreement No. 678763).

Contents

1 **Introduction** 1
  1.1   General 1
  1.2   Simulation Methods for Complex Electronic/Phononic
       Structure Materials 2
  1.3   Simulation Methods for Nanostructured Materials 3
  1.4   Brief Outlook 5
  References 6

2 **Boltzmann Transport Method for Electronic Transport
in Complex Bandstructure Materials** 9
  2.1   General 9
  2.2   The Boltzmann Transport Equation 9
  2.3   Thermoelectric Coefficients and the ZT Figure of Merit 15
  2.4   Carrier Scattering 21
  2.5   Ballistic Transport—The Landauer Approach 27
  2.6   Numerical Extraction of DOS (E), $v(E)$, and $\tau(E)$ from Arbitrary
       Bandstructures 28
  2.7   Conclusions 33
  References 34

3 **Monte Carlo Method for Electronic and Phononic Transport
in Nanostructured Thermoelectric Materials** 37
  3.1   General 37
  3.2   Monte Carlo Simulation Steps 38
  3.3   Nanostructuring and Boundary Scattering in Monte
       Carlo Simulations 44
  3.4   Thermoelectric Coefficients Extraction from Monte Carlo 47
  3.5   Phonon Monte Carlo Simulation Method Specifics 49
  3.6   Conclusions 56
  References 57

4   Non-Equilibrium Green's Function Method for Electronic
Transport in Nanostructured Thermoelectric Materials   59
   4.1   General   59
   4.2   The Non-Equilibrium Green's Function (NEGF) Method   60
   4.3   1D Numerical Implementation of NEGF   63
   4.4   Beyond 1D—Numerical Implementation of NEGF
        for Channels of Finite Cross Section   65
   4.5   Numerical Implementation of Electron-Phonon Scattering
        Within NEGF   68
   4.6   Recursive Green's Function Algorithm Applied in NEGF   72
   4.7   Ballistic to Diffusive, and Coherent to Incoherent Transport
        Regimes   73
   4.8   Simulation Procedure for the Extraction of the Thermoelectric
        Coefficients   74
   4.9   Simulation Example for Hierarchically Nanostructured
        Materials   76
   4.10 Conclusions   79
   References   79

5   Summary and Concluding Remarks   81
   References   84

Index   85

Symbols

σ	Electrical conductivity
S	Seebeck coefficient
T	Temperature
υ	Velocity
ν	Bandstructure velocity
υ_{d}	Drift velocity
F	Force
m	Mass
E_{field}	Electric field
f_0	Equilibrium distribution function
f	Distribution function
f_{S}	Symmetric distribution function
f_{A}	Antisymmetric distribution function
n	Carrier density
τ_{S}	Scattering time
p	Momentum
$\Im$	Flux
μ_0	Mobility
q	Electron charge
g	Density of states
Ξ	Transport distribution function
I	Current
V_{OC}	Open circuit voltage
κ	Thermal conductivity
κ_{e}	Electronic part of the thermal conductivity
κ_{l}	Lattice part of the thermal conductivity
κ_{bi}	Bipolar part of the thermal conductivity
κ_{tot}	Total thermal conductivity
E_{F}	Fermi energy
E_{C}	Band energy

η_{F}	Reduced Fermi level
Ω	Volume
ψ	Wavefunction
L_{D}	Screening length
N_{I}	Impurity density
h	Planck's constant
Q_{super}	Super-electron charge
Δ_{rms}	Roughness root-mean-square
$\langle E \rangle$	Average energy of the current flow
$\langle TOF \rangle$	Average time of flight
λ_{avg}	Average mean-free-path
H	Hamiltonian matrix
G	Green's function
G	Reciprocal unit cell length
G_{S}	Conductance driven by temperature gradient
Σ	Self-energy
ADP	Acoustic deformation potential
BTE	Boltzmann transport equation
DFPT	Density functional perturbation theory
DFT	Density functional theory
DOS	Density of states
FDM	Finite displacement method
IIS	Ionized impurity scattering
LA	Longitudinal acoustic
MC	Monte Carlo
MD	Molecular dynamics
MFP	Mean-free-path
NEGF	Non-equilibrium Green's function
ODP	Optical deformation potential
PF	Power factor
TA	Transverse acoustic

Chapter 1
Introduction

1.1 General

Thermoelectric materials convert heat through temperature gradients into electricity, and vice versa provide cooling capabilities once a potential difference is applied across them. The realization of complex bandstructure materials and their alloys, as well as nanostructured materials have revived the field of thermoelectric from decades of moderate activity, as they allow possibilities to largely improved performance. The conversion efficiency is quantified by the so-called ZT figure-of-merit, which depends on the electronic conductivity (σ), the Seebeck coefficient (S), the temperature T, and the thermal conductivity of the material (κ) as $ZT = \sigma S^2 T/\kappa$. The parameters that determine ZT are interconnected, with the electrical conductivity being reduced as the Seebeck increases, and vice versa.

A large number of new materials with complex crystal structures, including clathrates, skutterudites, half-Heuslers, oxides, silicides, chalcogenides, and selenides, to name a few, have recently been the subject of intense study from the experimental and theoretical point of view [1–6]. Their complex bandstructures provide possibilities to achieve high thermoelectric power factors (σS^2) by decoupling the electrical conductivity, σ, and the Seebeck coefficient, S, one of the key elements to increase ZT. In addition, their anharmonic bonds and nanostructured forms allow for drastic reductions in thermal conductivity, but also possibilities for power factor improvements [1, 7]. New design concepts such as the exploration of resonant states, energy filtering, modulation doping, and band alignment, allow favourable modifications of the electronic properties, and the nature of electron transport. Strongly anharmonic bonds, lattice imperfections, as well as strain fields, mass fluctuations, lattice mismatch due to disorder at the atomic scale, the introduction of second-phases and alloying [8, 9], significantly increase the scattering of short and medium phonon mean-free-paths (~<100 nm), ultimately leading to lower thermal conductivity, κ, [10], the second element that improves ZT.

N. Neophytou, *Theory and Simulation Methods for Electronic and Phononic Transport in Thermoelectric Materials*, SpringerBriefs in Physics, https://doi.org/10.1007/978-3-030-38681-8_1

The combination of complex bandstructure materials with rich features for the electronic/phononic bands, and nanostructuring, in particular hierarchical nanostructuring which can scatter phonons of different mean-free-paths across the phonon spectrum, has resulted in some of the highest ZT values [8], and emerges as the most promising direction for new generation thermoelectric materials.

1.2   Simulation Methods for Complex Electronic/Phononic Structure Materials

Alongside the advancement in experimental works, theory and simulation of electro-thermal transport properties of materials has also been rapidly advancing. Most of the theoretical and simulation work in the literature is focused on the identification and exploration of the properties of complex bandstructure materials. A variety of simulation software and techniques have been developed, or are in the process of being developed to improve the accuracy of these calculations. The most common simulation techniques for the thermoelectric properties of complex materials use ab initio techniques (DFT) [11–15], tight-binding techniques [16] or analytical techniques that provide the electronic/phonon dispersion relations of the materials. For electrons, more computationally demanding many-body perturbation methods, such as the GW approximation [17, 18] are employed when higher accuracy is required. Phonon spectra are often obtained by analyzing the forces associated with a systematic set of atomic displacements using the finite displacement method (FDM) [19, 20]. Alternative approaches include density functional perturbation theory (DFPT) [21, 22], where direct calculation of second order derivatives of the energy, provide inter-atomic force constants. Albeit less accurate than first principles approaches, classical potentials and Molecular Dynamics can also be used to extract phonon dispersions and other phonon transport properties [23, 24].

Once the dispersion relations are obtained, the semi-classical Boltzmann Transport Equation (BTE), (or the Landauer formalism in some cases), is used to extract the thermoelectric transport coefficients – typically in the relaxation time approximation and in conjunction with the rigid band approximation. Many software packages that are interfaced with DFT packages exist for this part of the calculation. For instance, *BoltzTraP* [25] and *BoltzWann* [26] are software that can be used to solve the semi-classical Boltzmann transport equation for electrons, using as inputs calculated electron bandstructures. *Phono3py* [27] and *ShengBTE* [28] are software used for solving the Boltzmann transport equation for phonons, using as main inputs the 2nd and 3rd order interatomic force constants. Recent progress in linear scaling DFT software [29, 30] enabled DFT calculations on large systems, and could be used for generation of more accurate bandstructures for highly doped structures and alloys, beyond the rigid band approximation as well. Other sophisticated recent approaches use ab initio extracted electron-phonon scattering rates such as the Electron-Phonon-Wannier

(*EPW*) software [31], which allows electronic transport considerations beyond the simplified constant relaxation time approximation and beyond deformation potential theory. Similarly, when it comes to phonon transport, molecular dynamics approaches [32] that can treat anharmonicity to all orders, or techniques that involve the exact solution of the BTE, are employed in order to go beyond the phonon-phonon relaxation time approximation [33, 34]. In addition, in order to accelerate the discovery of promising thermoelectric materials in terms of the power factor from the myriad of potential alloys and compounds, high throughput screening [35, 36] of DFT calculated material properties is often combined with machine learning techniques [35, 37]. Descriptors that point towards high power factors, low thermal conductance, and thermal, mechanical, and chemical stability, are being developed. The ability to decouple σ and S is also used as an additional efficient measure for screening [38, 39]. Although DFT-based calculations are commonly employed, in the majority of studies, the solutions to the BTE assume the constant relaxation time approximation, despite the fact that we know that the scattering mechanisms are not only energy, but momentum, and band dependent. This is due to the vast computational costs in treating energy dependent scattering mechanisms properly. Thus, in Chap. 2 we address how to construct a simulation setup to properly account for energy dependences.

1.3   Simulation Methods for Nanostructured Materials

With the emergence of nanostructuring, serious simulation efforts related to electronic/thermal transport in nanostructured materials have also emerged. Initial efforts were focused on thermal conductivity reduction in such materials, as the multiple internal inclusions can provide phonon scattering centres to reduce the thermal conductivity. A popular tool to compute the thermal conductivity in nanostructures is Molecular Dynamics (MD). MD can capture phonon transport in nanostructured materials at the atomistic level including anharmonic effects (and wave effects), and can treat large domains with several hundred thousand atoms, something not computationally affordable with first principles methods. MD has been extensively used in understanding the role of surfaces, boundaries, disorder, discontinuities, and interfaces in nanostructures and superlattices, cases where the thermal conductivity can drop below the amorphous limit [40–44].

In classical MD the forces on all atoms, defined by an interatomic potential, are evaluated and used to solve Newton's equations of motion to determine how the atoms evolve. Choosing the best available potential and understanding its limitations is therefore crucial, and not all potentials agree on thermal conductivity predictions. Interatomic potentials are generally based on simple functional forms with adjustable parameters, but can have varying degrees of complexity. The problem of accurately mimicking true energy surfaces is far from trivial, and the parameters are often chosen such that the empirical potential matches results obtained with first principles calculations or experimental data [45]. Potentials vary in terms of the nature of the bonding described (covalent, polar covalent, ionic, metallic, hydrogen, van der Waals), whether short- and/or long-range interactions are considered, and the number of atoms that interact with each other (pair potentials versus many-body potentials) [45, 46]. Of interest to thermoelectric materials, the effect of imperfections in lowering the mean-free-path of materials is more noticeable at low to medium temperatures. Phonon-phonon scattering is weak in this regime and defect scattering plays a more significant role [47]. However, strictly speaking, classical molecular dynamics is only applicable to solids above the Debye temperature, below which quantum effects are more prevalent. In some cases, for studies below the Debye temperature, interatomic potentials designed to approximate quantum behaviour can be used, as well as quantum corrections [43].

Thermal transport in MD, and related quantities (such as phonon relaxation times and velocities), are generally obtained by computing thermal conductivity using one of two major approaches: (1) equilibrium, via the Green–Kubo formalism [48, 49], and (2) non-equilibrium or direct methods, such as the Müller-Plathe [50]. Each has advantages and disadvantages, and the method chosen strongly depends on the problem of interest [51]. In general, the equilibrium approach is preferable for bulk phase simulations and the direct method is best for finite structures [43]. Direct methods rely on disturbing a system and measuring its response, while in the equilibrium MD approach the response is computed from the averaging of small, local deviations from equilibrium that occur in the simulation time domain. Finally, MD can elucidate the phonon transport details in micrometer long structures, but it cannot computationally scale to large material domains in both the width/length and even depth, which would require the consideration of multi-million atoms. Such large domains are needed in order to understand heat transport in hierarchical nanostructured materials, where defects are encountered on the atomistic, nanoscale and macroscale. For this, continuum models are needed, and one of them is the use of Monte Carlo simulations, which we describe in Chap. 3. Monte Carlo simulations trace a large number of classical particles under certain scattering rules, and transport results are extracted after enough statistics are gathered. In such they provide flexibility in simulation domains, geometrical complexities, and scattering details.

Although nanostructuring can reduce the electronic conductivity, it can also provide possibilities for manipulating the charge carrier transport in ways that improve the Seebeck coefficient (for example through energy filtering) and/or the electrical conductivity (for example through modulation doping), and thus, the power factor as well. Simulation techniques for nanostructured thermoelectric materials range from analytical models, semi-classical Monte Carlo simulations for electronic transport, to fully quantum mechanical Non-Equilibrium Green's Function simulations. Each method has its advantages and limitations, regarding accuracy and computational cost, as well as their ability to merge length scales and different physics. They can, however, provide the means for power factor optimization in the presence of thermal conductivity reduction which is beneficial to ZT, and literature reports on how to achieve this are beginning to appear [52]. These techniques are well-established and traditionally employed for transistor devices, but only recently are beginning to emerge as useful tools for understanding electronic transport in nanostructured thermoelectric materials. Chapters 3 and 4 deal with the simulation setup of these methods.

1.4   Brief Outlook

Chapter 2 presents the Boltzmann Transport formalism, which leads to the thermo-electric coefficients (electrical conductivity, Seebeck coefficient, and the electronic part of the thermal conductivity), and then it continues to the derivation of the ZT figure of merit from simple circuit analysis considerations. A discussion on scattering mechanisms follows, and then an elaborate description on the numerical implementation of the extraction of the thermoelectric coefficients in 2D and 3D for arbitrary bandstructure materials—crucially including all energy, momentum and band dependencies for all relevant scattering relaxations times. While theoretical discussions are included, the focus is placed on the numerical implementation of the process, rather than involved theoretical description, which can be found elsewhere.

Chapters 3 and 4 describe the implementations of 'real-space' methods, i.e. those that explore thermoelectric coefficients in nanostructured materials by explicitly taking into account the complexities in the geometries. In these techniques the numerical implementation begins with the definition of the geometrical features of the nanostructured material, and then the flow of particles is considered. Chapter 3 describes the semi-classical Monte Carlo technique, which treats electrons (or phonons) as particles that flow through the material under consideration, essentially following Newton's laws in a raytracing manner and under certain scattering rules. Both electronic and phonon Monte Carlo simulation method details are presented.

Chapter 4 then moves on to describe the Non-Equilibrium Green's Function (NEGF) implementation of thermoelectric material simulations, which considers the full quantum mechanical wave nature of electrons. Importantly, I present the implementation of the electron-phonon scattering, and I explain why this constitutes an essential aspect of thermoelectric transport in modern thermoelectric materials.

Finally, Chap. 5 provides some concluding remarks considering the simulations and attempts an outlook into the role that such simulation methods can play in the design of the next generation thermoelectric materials. I discuss this briefly in the context of materials screening, machine learning, and nanostructuring concepts for large power factors to complement the low thermal conductivity efforts.

References

1. Beretta, D., Neophytou, N., Hodges, J.M., Kanatzidis, M., Narducci, D., Martin-Gonzalez, M., Beekman, M., Balke, B., Cerretti, G., Tremel, W., Zevalkink, A., Hofmann, A.I., Müller, C., Dörling, B., Campoy-Quiles, M. & Caironia, M.: Thermoelectrics: from history, a window to the future. Mater. Sci. Eng. R **138**, 210–255 (2019)
2. Lei, C., Burton, M.R., Nandhakumar, I.S.: Facile production of thermoelectric bismuth telluride thick films in the presence of polyvinyl alcohol. Phys. Chem. Chem. Phys. **18**(21), 14164–14167 (2016)
3. Mele, P., Narducci, D., Ohta, M., Biswas, K., Morante, J.R., Saini, S., Endo, T. (eds.): Thermoelectric Thin Films: Materials and Devices. Springer (2019)
4. Koumoto, K., Mori, T.: Thermoelectric Nanomaterials. Springer (2015)
5. Liu, W., Hu, J., Zhang, S., Deng, M., Han, C.G., Liu, Y.: New trends, strategies and opportunities in thermoelectric materials: a perspective. Mater. Today Phys. **1**, 50–60 (2017)
6. Goldsmid, H.J.: Introduction to Thermoelectricity, pp. 9–24. Heidelberg (2016)
7. Neophytou, N., Zianni, X., Kosina, H., Frabboni, S., Lorenzi, B., Narducci, D.: Nanotechnology **24**, 205402 (2013)
8. Biswas, K., He, J., Blum, I.D., Wu, C.-I., Hogan, T.P., Seidman, D.N., Dravid, V.P., Kanatzidis, M.G.: High-performance bulk thermoelectrics with all-scale hierarchical architectures. Nature **489**, 414–418 (2012)
9. Hattori, K., Miyazaki, H., Yoshida, K., Inukai, M., Nishino, Y.: Direct observation of the electronic structure in thermoelectric half-heusler alloys zr1–x m x nisn (m = y and nb). J. App. Phys.**117**, 205102 (2015)
10. Perez-Taborda, J.A., Rojo, M.M., Maiz, J., Neophytou, N., Martin-Gonzalez, M.: Ultra-low thermal conductivities in large-area Si-Ge nanomeshes for thermoelectric applications. Sci. Rep. **6**, 32778 (2016)
11. Yokobori, T., Okawa, M., Konishi, K., Takei, R., Katayama, K., Oozono, S., Shinmura, T., Okuda, T., Wadati, H., Sakai, E., Ono, K.: Electronic structure of the hole-doped delafossite oxides $CuCr_{1-x}Mg_xO_2$. Phys. Rev. B **87**, 195124 (2013)
12. Lu, X., Morelli, D.T., Wang, Y., Lai, W., Xia, Y., Ozolins, V.: Phase stability, crystal structure, and thermoelectric properties of $Cu_{12}Sb_4S_{13-x}Se_x$ solid solutions. Chem. Mater. **28**(6), 1781–1786 (2016)
13. Lu, X., Morelli, D.T., Xia, Y., Ozolins, V.: Increasing the thermoelectric figure of merit of tetrahedrites by Co-doping with nickel and zinc. Chem. Mater. **27**(2), 408–413 (2015)
14. Xi, L., Zhang, Y.B., Shi, X.Y., Yang, J., Shi, X., Chen, L.D., Zhang, W., Yang, J., Singh, D.J.: Chemical bonding, conductive network, and thermoelectric performance of the ternary semiconductors Cu_2SnX_3 (X = Se, S) from first principles. Phys. Rev. B **86**(15), 155201 (2012)
15. Flage-Larsen, E., Diplas, S., Prytz, Ø., Toberer, E.S., May, A.F.: Valence band study of thermo-electric Zintl-phase $SrZn_2Sb_2$ and $YbZn_2Sb_2$: X-ray photoelectron spectroscopy and density functional theory. Phys. Rev. B **81**(20), 205204 (2010)
16. Neophytou, N., Wagner, M., Kosina, H., Selberherr, S.: Analysis of thermoelectric properties of scaled silicon nanowires using an atomistic tight-binding model. J. Electr. Mater. **39**(9), 1902–1908 (2010)
17. van Schilfgaarde, M., Kotani, T., Faleev, S.: Quasiparticle self-consistent g w theory. Phys. Rev. Lett. **96**(22), 226402 (2006)
18. Cheng, L., Liu, H.J., Zhang, J., Wei, J., Liang, J.H., Jiang, P.H., Fan, D.D., Sun, L., Shi, J.: High thermoelectric performance of the distorted bismuth (110) layer. Phys. Chem. Chem. Phys. **18**(26), 17373–17379 (2016)
19. Kresse, G., Furthmüller, J., Hafner, J.: Ab initio force constant approach to phonon dispersion relations of diamond and graphite. EPL (Europhys. Lett.) **32**(9), 729 (1995)
20. Parlinski, K., Li, Z.Q., Kawazoe, Y.: Parlinski, Li, and Kawazoe reply. Phys. Rev. Lett. **81**(15), 3298 (1998)
21. Giannozzi, P., De Gironcoli, S., Pavone, P., Baroni, S.: Ab initio calculation of phonon dispersions in semiconductors. Phys. Rev. B **43**(9), 7231 (1991)

22. Gonze, X., Beuken, J.M., Caracas, R., Detraux, F., Fuchs, M., Rignanese, G.M., Sindic, L., Verstraete, M., Zerah, G., Jollet, F., Roy, A., Mikami, M., Ghosez, P., Raty, J.Y., Allan, D.C., Torrent, M.: First-principles computation of material properties: the ABINIT software project. Comput. Mater. Sci. **25**(3), 478–492 (2002)
23. Kong, L.T.: Phonon dispersion measured directly from molecular dynamics simulations. Comput. Phys. Commun. **182**(10), 2201–2207 (2011)
24. Kang, J., Wang, L.W.: First-principles Green-Kubo method for thermal conductivity calculations. Phys. Rev. B **96**(2), 020302 (2017)
25. Madsen, G.K., Singh, D.J.: BoltzTraP. A code for calculating band-structure dependent quantities. Comput. Phys. Commun. **175**(1), 67–71 (2006)
26. Pizzi, G., Volja, D., Kozinsky, B., Fornari, M., Marzari, N.: BoltzWann: A code for the evaluation of thermoelectric and electronic transport properties with a maximally-localized Wannier functions basis. Comput. Phys. Commun. **185**(1), 422–429 (2014)
27. Togo, A., Chaput, L., Tanaka, I.: Distributions of phonon lifetimes in Brillouin zones. Phys. Rev. B **91**(9), 094306 (2015)
28. Li, W., Carrete, J., Katcho, N.A., Mingo, N.: ShengBTE: A solver of the Boltzmann transport equation for phonons. Comput. Phys. Commun. **185**(6), 1747–1758 (2014)
29. Bowler, D.R., Miyazaki, T.: Calculations for millions of atoms with density functional theory: linear scaling shows its potential. J. Phys. Condens. Matter **22**(7), 074207(2010)
30. Skylaris, C.K., Haynes, P.D., Mostofi, A.A., Payne, M.C.: Introducing ONETEP: linear-scaling density functional simulations on parallel computers. J. Chem. Phys. **122**(8), 084119 (2005)
31. Poncé, S., Margine, E.R., Verdi, C., Giustino, F.: EPW: electron–phonon coupling, transport and superconducting properties using maximally localized Wannier functions. Comput. Phy. Commun. **209**, 116–133 (2016)
32. Hellman, O., Broido, D.A.: Phonon thermal transport in Bi_2Te_3 from first principles. Phys. Rev. B **90**(13), 134309 (2014)
33. Chiloyan, V., Huberman, S., Ding, Z., Mendoza, J., Maznev, A.A., Nelson, K.., Chen, G.: Micro/nanoscale thermal transport by phonons beyond the relaxation time approximation: Green's function with the full scattering matrix. arXiv:1711.07151 (2017)
34. Broido, D.A., Malorny, M., Birner, G., Mingo, N., Stewart, D.A.: Intrinsic lattice thermal conductivity of semiconductors from first principles. Appl. Phys. Lett. **91**(23), 231922 (2007)
35. Carrete, J., Li, W., Mingo, N., Wang, S., Curtarolo, S.: Finding unprecedentedly low-thermal-conductivity half-Heusler semiconductors via high-throughput materials modeling. Phys. Rev. X **4**(1), 011019 (2014)
36. Zhu, H., Hautier, G., Aydemir, U., Gibbs, Z.M., Li, G., Bajaj, S., Pöhls, J.-H., Broberg, D., Chen, W., Jain, A., White, M.A.: Computational and experimental investigation of TmAgTe2 and XYZ2 compounds, a new group of thermoelectric materials identified by first-principles high-throughput screening. J. Mater. Chem. C **3**(40), 10554–10565 (2015)
37. Oliynyk, A.O., Antono, E., Sparks, T.D., Ghadbeigi, L., Gaultois, M.W., Meredig, B., Mar, A.: High-throughput machine-learning-driven synthesis of full-Heusler compounds. Chem. Mater. **28**(20), 7324–7331 (2016)
38. Xing, G., Sun, J., Li, Y., Fan, X., Zheng, W., Singh, D.J.: Electronic fitness function for screening semiconductors as thermoelectric materials. Phys. Rev. Mater. **1**(6), 065405 (2017)
39. Chen, X., Parker, D., Singh, D..: Importance of non-parabolic band effects in the thermoelectric properties of semiconductors. Sci. Rep. **3**, 3168 (2013)
40. Li, D., McGaughey, A.J.: Phonon dynamics at surfaces and interfaces and its implications in energy transport in nanostructured materials—an opinion paper. Nanoscale Microscale Thermophys. Eng. **19**(2), 166–182 (2015)
41. Xu, Z.: Heat transport in low-dimensional materials: a review and perspective. Theor. Appl. Mech. Lett. **6**(3), 113–121 (2016)
42. Katz, H.E., Poehler, T.O. (eds.): Innovative Thermoelectric Materials: Polymer, Nanostructure and Composite Thermoelectrics. World Scientific (2016)
43. McGaughey, A.J., Kaviany, M: Phonon transport in molecular dynamics simulations: formulation and thermal conductivity prediction. Adv. Heat Transf. **39**, 169–255 (2006)

44. de Sousa Oliveira, L., Neophytou, N.: Large-scale molecular dynamics investigation of geometrical features in nanoporous Si. Phys. Rev. B **100**, 035409 (2019)
45. Meller, J. et al.: Molecular dynamics. ELS (2001)
46. LeSar, R.: Introduction to Computational Materials Science: Fundamentals to Applications. Cambridge University Press (2013)
47. Tian, Z., Lee, S., Chen. G.: A comprehensive review of heat transfer in thermoelectric materials and devices. Ann. Rev. Heat Transf. **17**, 425–483 2014
48. Green, M.S.: Markoff random processes and the statistical mechanics of time-dependent phenomena. II. Irreversible processes in fluids. J. Chem. Phys. **22**(3), 398–413 1954
49. Kubo, R.: Statistical-mechanical theory of irreversible processes. I. general theory and simple applications to magnetic and conduction problems. J. Phys. Soc. Jpn. **12**(6), 570–586 (1957)
50. Müller-Plathe, F.: A simple nonequilibrium molecular dynamics method for calculating the thermal conductivity. J. Chem. Phys. **106**(14), 6082–6085 (1997)
51. Sellan, D.P., Landry, E.S., Turney, J.E., McGaughey, A.J.H., Amon, C.H.: Size effects in molecular dynamics thermal conductivity predictions. Phy. Rev. B **81**(21), 214305 (2010)
52. Vargiamidis, V., & Neophytou, N.: Hierarchical nanostructuring approaches for thermoelectric materials with high power factors. Physical Review B, 99(4), 045405, (2019)

Chapter 2
Boltzmann Transport Method for Electronic Transport in Complex Bandstructure Materials

2.1 General

The semiclassical Boltzmann Transport Equation (BTE) is key in theoretical calculations of thermoelectric properties, both for electrons and phonons. The chapter will cover the physics of semiclassical transport methods (mainly the Boltzmann transport in the relaxation time approximation and briefly the Landauer approach). The chapter continuous in deriving the thermoelectric coefficients from a circuit analysis point of view, and relates those to the BTE. Further on, it discusses the scattering physics for different scattering mechanisms for electronic transport, namely electron-phonon scattering, ionized impurity scattering, alloy scattering, and boundary scattering. Afterwards, the chapter presents a numerical scheme that allows for the consideration of the energy dependences of the scattering mechanisms coupled with arbitrary bandstructure shapes that characterize the new generation complex bands thermoelectric materials. Such scheme can take the calculations beyond the simplistic constant relaxation time approximation within the BTE, and allow more accuracy in calculations, more design opportunities for high power factor material designs, as well as more reliable identification of promising thermoelectric materials through materials screening methods.

2.2 The Boltzmann Transport Equation

The Boltzmann Transport Equation (BTE) is the most common theoretical formalism employed to the studies of thermoelectric materials. The BTE is essentially an equation for the distribution function of particles in position and velocity/momentum spaces. The particles obey Newton's laws of motion, and particle positions and velocities are all accounted usually in 3D, unless we are dealing with low-dimensional materials. Thus, the BTE becomes a 6-dimensional problem (three dimensions for

N. Neophytou, *Theory and Simulation Methods for Electronic and Phononic Transport in Thermoelectric Materials*, SpringerBriefs in Physics, https://doi.org/10.1007/978-3-030-38681-8_2

velocities, and three for positions). In typical thermoelectric studies, the electronic structure of a material is computed using ab initio Density Functional Theory (DFT) calculations usually on a sparse grid, then interpolated using appropriate methods, and them employed within the BTE to obtain the thermoelectric coefficients, most commonly in the relaxation time approximation [1, 2]. For the purposes of this brief, I do not go deeply into the theory of BTE, DFT, or the extraction of the relaxations times. I provide sufficient details to allow one to understand and be able to implement these computational methods and then extract the thermoelectric coefficients. Detailed derivations are found elsewhere in numerous places and the interested reader is directed to those references [3–5]. In particular, I provide numerical details on how to account for the fully numerical dependence of the scattering rates, which will be a function of energy, momentum, and band. As I will also explain, such treatment has important implications for the thermoelectric coefficients of complex bandstructure materials, and different performance conclusions can be reached if this complexity is taken into account, rather than if simplified (but computationally efficient), constant relaxation scattering times are utilized for the scattering mechanisms.

Pictorially, the motion of particles can be described in a 6-dimensional space, with three coordinates describing the position, and the other three the velocity of particles, as depicted in Fig. 2.1. The collective viewpoint, is that one can form the distribution function $f(x, v, t)$, which is a function that describes the positions and momenta/velocities of all particles i.e. how many electrons at time t, are located at position x_i, and have velocity v_i. Once this function is known, then the current can be computed by multiplying the unit charge of each electron by its velocity and summing them up as:

$$J = \frac{1}{L} \sum_{v_x} (-q) v_x f(v_x) \tag{2.1}$$

Above I consider the one-dimensional (1D) problem, as it is easier to derive the quantities needed, but easily one can generalize to 3D, as I do at the end of the derivations. The distribution function in general changes in time due to the application of a driving force, F, for example the application of a potential difference, or a temperature difference. Dissipative and randomizing processes are also present (such as scattering), which tend to bring the distribution to equilibrium, i.e. after the removal

Fig. 2.1 The motion of particles in position and velocity spaces

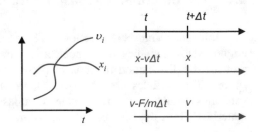

of the applied driving force. How the distribution function changes in time, simply follows from the previous time attributes by adding the distance travelled to the old position of the particles, and by adding the velocity change to the old velocity of the particles, as:

$$f(x, \upsilon, t + \Delta t) = f\left(x - \upsilon \Delta t, \upsilon - \frac{F}{m}\Delta t, t\right) \quad (2.2)$$

Above F is the force applied on the particles (we will consider electrons in this case, but the BTE is in general applied to mass particles, phonons etc.). m is the mass of the particles, in the case of electrons it is their effective mass. For small time steps, using Taylor series expansion, and the mathematical chain rule, we can write the distribution function as:

$$
\begin{aligned}
f(x, \upsilon, t + \Delta t) &= f(x, \upsilon, t) + \frac{\partial f}{\partial t}\Delta t \\
&= f(x, \upsilon, t) + \left(\frac{\partial f}{\partial x}\frac{\partial x}{\partial t} + \frac{\partial f}{\partial \upsilon}\frac{\partial \upsilon}{\partial t}\right)\Delta t \\
&= f(x, \upsilon, t) + \left(\frac{\partial f}{\partial x}(-\upsilon) + \frac{\partial f}{\partial \upsilon}\left(-\frac{F}{m}\right)\right)\Delta t \quad (2.3)
\end{aligned}
$$

By bringing everything to the left hand side, we reach to the so-called collisionless BTE:

$$\frac{\partial f}{\partial t} + \upsilon \nabla_r f + \frac{F}{m}\nabla_\upsilon f = 0 \quad (2.4)$$

The second and third terms of the equation above refer to diffusion and drift processes, respectively. However, carriers interact with the environment and undergo scattering. If an excitation is applied on the system and then removed, the system will return to equilibrium within a certain time, the so-called relaxation time. Thus, the BTE needs to be modified to include the so-called 'scattering operator', usually denoted by S_{op}.

$$\frac{\partial f}{\partial t} + \upsilon \nabla_r f - qE_{field}\nabla_p f = \frac{\partial f}{\partial t}|_{collisions} = S_{op}f = -\frac{f - f_0}{\tau_s} \quad (2.5)$$

Notice that above we replaced the derivative in terms of the velocity with the derivative in terms of the momentum, which removes the particle mass from the equation, and replaced the force on the particles with the electric field $-qE_{field}$ (the relevant force for charged particles). Using the carrier's momentum is a commonly used notation in the literature, but as we will see later on after the derivation of the BTE, we shift back to velocities, which is a more convenient way to compute the transport properties of complex bandstructure thermoelectric materials. The term $\frac{\partial f}{\partial t}|_{collisions} = -\frac{f - f_0}{\tau_s}$ is the simple approximation that we commonly make for the scattering operator, namely, indicating that the rate at which the distribution function relaxes back to its equilibrium state depends on how far off equilibrium it is positioned, and on a characteristic time constant, the so-called scattering relaxation time. This is very similar to how we describe other physical processes that tend to relax back to equilibrium within a characteristic time, as for example the recombination-generation statistics (R-G) for the change in the minority carrier concentration, dn/dt = $-\Delta n/\tau$ [6]. The solution to that equation is $\Delta n(t)= \Delta n(0)\exp(-t/\tau)$, where the initial concentration $\Delta n(0)$ at the time where the excitation is removed is the boundary condition. This means that if the system is left to relax after an excitation, it will do so exponentially with a characteristic relaxation time. We assume the same relaxation time as the simplest form of the collision operator. This is called the *relaxation time approximation* (RTA). The details of how τ_s is defined will be presented below.

Assuming steady state such that $\frac{\partial f}{\partial t} = 0$, and assuming a long uniform system such that no carrier variations exist, then $\nabla_r n = 0$ (meaning that we ignore any diffusion processes), and the BTE is simplified to:

$$f \simeq f_0 + q\tau_s E_{\text{field}} \nabla_p f \tag{2.6}$$

The result in (2.6) above is important, and essentially it says that the new distribution which describes the system after an electric field being applied, is the same as the old distribution f_0, plus an additional term that depends on the derivative of the distribution with respect to momentum. In the case when that term being small, also referred to as 'linear response', or the formalism as 'linearized Boltzmann transport', we can use Taylor series expansion backwards, having:

$$f(p_x) = f_0(p_x + q\tau_s E_{\text{field}}) \tag{2.7}$$

This again indicates as earlier, that the new distribution function (of momenta in the system), is the old one shifted by a (small) momentum term which depends on the electric field and the relaxation time.

Pictorially, this is described in Fig. 2.2. A typical distribution function has the shape of a Maxwellian distribution. Under equilibrium conditions, it contains equal number of carriers with positive and negative momenta, indicated by the symmetric around zero **blue line** in Fig. 2.2a. When the argument of the function is shifted, the *displaced* Maxwellian is indicated by the **orange line**, indicating a difference in the numbers of carriers with positive versus negative momenta. This is created upon the application of an electric field in the negative direction, and as a consequence of this imbalance current will flow. Mathematically, since the symmetric distribution does not provide any current, a simple rearrangement of (2.6) to:

$$f - f_0 \equiv f_A \simeq q\tau_s E_{\text{field}} \nabla_p f \tag{2.8}$$

denotes the so-called antisymmetric part of the distribution function, i.e. the deviation from symmetric conditions. Its shape is as shown in Fig. 2.2b, which indicates

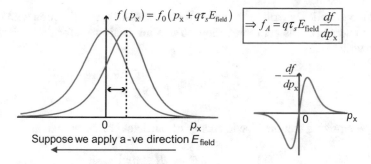

Fig. 2.2 a The equilibrium distribution function (blue line) and the 'displaced' distribution function after the application of an electric field (orange line). **b** The derivative of the distribution function with respect to the momentum, indicating a surplus of carriers with positive momenta, and a deficit of carriers with negative momenta

depletion of carriers with negative momenta (or velocities), and an excess of carries with positive momenta (or velocities). This can be clearly observed if one plots the derivative of the distribution function from (2.6). Thus, it is the antisymmetric part of the distribution function which contributes to current, and the symmetric part does not, as the two opposite fluxes counter-balance each other. Mathematically, the net flux $\Im$ can be expressed as the summation of the symmetric f_S and antisymmetric distributions f_A as:

$$\Im = \frac{1}{L}\sum_k v_x f = \frac{1}{L}\sum_k v_x(f_S + f_A)$$

$$= \frac{1}{L}\sum_k v_x f_S + \frac{1}{L}\sum_k v_x f_A$$

$$= 0 + \frac{1}{L}\sum_k v_x f_A$$

$$\Rightarrow \Im = \frac{1}{L}\sum_k v_x f_A \tag{2.9}$$

Thus, the net flux ends up being only the summation of the antisymmetric part of the distribution, weighted by the velocities of the particles. Notice that I have use a normalization constant $1/L$ as I have assumed an 1D channel.

Substituting (2.8) for the antisymmetric part of the distribution function into (2.9) for the flux, we obtain:

$$\Im = \frac{1}{L}\sum_k v_x q\tau_s E_{field}\frac{\partial f}{\partial p_x} \tag{2.10}$$

where I have explicitly written the derivative of the distribution function with respect to the momentum. It is convenient now to use the chain rule for the derivative in order to write:

$$\Im = \frac{1}{L}\sum_k \upsilon_x q \tau_s E_{\text{field}} \frac{\partial f}{\partial E}\frac{\partial E}{\partial p_x} \tag{2.11}$$

and by recognizing that the kinetic energy of particles can be written as:

$$E = \frac{1}{2}m\upsilon_x^2 \Rightarrow \frac{\partial E}{\partial p_x} = \frac{\partial\left(\frac{1}{2}m\upsilon_x^2\right)}{\partial(m\upsilon_x)} = \upsilon_x \tag{2.12}$$

the flux can be written as:

$$\Im = \frac{1}{L}\sum_k q\tau_s\upsilon_x^2 E_{\text{field}}\frac{\partial f}{\partial E} \tag{2.13}$$

To obtain the current density it is sufficient to multiply the flux with the electronic charge $-q$, such as:

$$J = (-q)\Im = q^2 E_{\text{field}}\frac{1}{L}\sum_k \tau_s\upsilon_x^2\left(-\frac{\partial f}{\partial E}\right) \tag{2.14}$$

From here, we notice that the summation over all k-space states is equivalent to an integral over the density-of-states (g) in energy, as:

$$J = q^2 E_{\text{field}}\int_E \tau_s\upsilon_x^2 g\left(-\frac{\partial f}{\partial E}\right)dE \tag{2.15}$$

and from this, the conductivity, which is defined as $\sigma = J/E_{\text{field}}$, can be extracted as:

$$\sigma = q^2\int_E \tau_s\upsilon_x^2 g\left(-\frac{\partial f}{\partial E}\right)dE \tag{2.16}$$

We note here that the distribution function f, under low-field transport, can be assumed to be the equilibrium distribution of carriers, following Boltzmann statistics, or Fermi-Dirac statistics. For the purposes of thermoelectric transport, in which case the materials are highly doped and the Fermi level is usually around the band edge, we employ the most appropriate Fermi-Dirac statistics, and thus the Fermi distribution as the equilibrium distribution function.

Following the result of (2.16), there are two points that I would like to make. The first one, is the nature of transport. We have learned in the drift-diffusion formulation that the current and conductivity can be expressed as $J_{dd} = qn\upsilon_d$, and $\sigma = qn\mu_0$, which is interpreted as if the entire change density n is travelling with the rather slow, drift velocity v_d, which is interlinked with the carrier mobility. In the BTE viewpoint, however, only a portion of the charge density is participating to transport, namely only the carriers having energies around the Fermi level (picked up by its derivative as shown in Fig. 2.2), quantitatively defined by the $k_B T$ depended broadening. Those carriers travel with their bandstructure velocity, which is overall higher compared to the drift velocity, but they scatter on average every relaxation time τ_s, whose strength limits the conductivity, randomizing the motion of electrons, allowing the slower drift to emerge.

The second important point, is the quantity:

$$\Xi(E) = \tau_s(E)v_x^2(E)g(E) \tag{2.17}$$

which appears under the integral of (2.16) for the conductivity. This is the so-called 'transport distribution function' (TDF), or the 'transport function' [7]. Although under this simplified 1D derivation the TDF is a function of v_x^2, it should be understood that in 3D it is a tensor quantity, effectively having $v_i v_j$ components, where i and j being x-, y-, and z-coordinates. This quantity turns out to be crucial in understanding electronic and thermoelectric transport in advanced materials. Notice that, importantly, all quantities that define it, the scattering (momentum) relaxation times, the bandstructure velocities, and the density of states, are all energy dependent. In principle, they are energy, momentum, and band dependent for complex bandstructure materials. It is interesting to elaborate first on the energy dependence of the transport distribution function, which is an integrated part of the thermoelectric coefficients. For this we make some simple assumptions for the energy dependence of the scattering times. Assuming simple isotropic acoustic phonon scattering, we obtain a scattering rate proportional to the density of states (we will discuss in more detail the scattering rates below), thus relaxation times inversely proportional to the density of states. The energy dependence of the scattering times and density of states then cancels out, and the transport distribution function becomes proportional to the bandstructure velocity squared. As the velocity has an $E^{1/2}$ proportionality, the TDF ends up being linear with energy, with the slope being inversely proportional to the (conductivity) effective mass of the band under consideration, i.e. $\Xi(E) \propto \frac{E}{m_C^*}$ [8]. Notice that the assumption of a constant relaxation time, results in a different energy dependence of the TDF, namely in 3D TDF $\sim E^{3/2}$, which has consequences in determining the TE coefficients of complex materials [9, 10].

2.3 Thermoelectric Coefficients and the ZT Figure of Merit

What we have derived up to now, is the electronic conductivity of the material based on the BTE, when an electric field (from a voltage difference) is applied across it. To extract the thermopower (or Seebeck coefficient), we need to extract the conductivity in the case where a temperature gradient applied across the material constitutes the driving force for the current. Above, the application of a potential voltage ΔV across the material results in a split between the electrochemical potentials of the contacts across the material. For transport purposes, this split is translated as the derivative of the Fermi distribution with energy in the BTE, df/dE, constituting the driving force for the current and the energy window in which it is active. In a similar manner, when the driving force is the temperature difference of the Fermi distributions of the two contacts, the corresponding conductivity follows a one-to-one correspondence with (2.16) as:

$$\sigma_S = q^2 \int_E \tau_s v_x^2 g \left(-\frac{\partial f}{\partial T} \right) dE \tag{2.18}$$

It can be shown in a straight-forward way that:

$$\left(-\frac{\partial f}{\partial T} \right) = \left(-\frac{\partial f}{\partial E} \right) \frac{(E - E_F)}{qT} \tag{2.19}$$

where the electronic charge q is inserted to convert the energies from electron-volts (eV) into Joules. When (2.19) is inserted into (2.18), after some simplification we reach:

$$\sigma_S = \frac{q}{T} \int_E \Xi(E)(E - E_F) \left(-\frac{\partial f}{\partial E} \right) dE \tag{2.20}$$

where $\Xi(E)$ is given as earlier by (2.17). Thus, the conductivity in the case where the driving forces behind the current are an applied voltage, or an applied temperature difference across the material, both involve the energy derivatives of the distribution function (Fermi distribution) and the transport distribution function, but in the latter case the product of these quantities is weighted by the energy. Pictorially the two cases are described by the schematics in Fig. 2.3, **first and second rows**, respectively. In the first case, the application of a voltage difference results in a shift of the E_F in the right contact, and the difference between the left/right distributions, $(f_1 - f_2)$, allows states around E_F, both above and below, to participate in transport. In the applied temperature difference case, however, the Fermi distribution in the left contact is broadened by a higher temperature. The transport is determined by the new shape of the df/dT, which now has a different sign for carriers above the Fermi level compared to carriers below the Fermi level. Essentially, this means that the corresponding conductivity is determined by the difference of the fluxes above and below the E_F, essentially by the energy asymmetry in transport above and below the Fermi energy level (rather than the asymmetry in momentum). This description finds several different manifestations in the literature, encountered in the derivative or slope of the density-of-states (DOS) around the E_F, the asymmetry in carrier scattering and mean-free-paths around E_F, etc. Strictly speaking, though, it is the

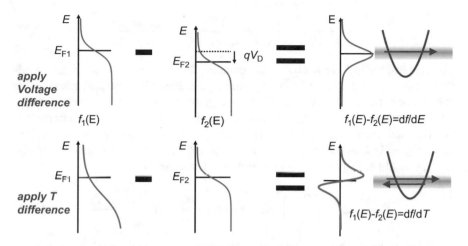

Fig. 2.3 The change in the Fermi distributions between the left and right contacts of a material, which drives the current. First row: the case of an applied voltage difference. Second row: the case of an applied temperature difference

asymmetry of the transport distribution function that determines the difference in the fluxes, as it is that quantity that is weighted by the derivative of the fermi function and the energy term in (2.20).

To determine the efficiency of the thermoelectric material, and the *ZT* figure of merit, we turn to circuit analysis. An elaborate discussion can be found in the book by Datta [11] as well as other resources, so here I only provide a basic derivation following [11]. From a circuit point of view, the total current density under the application of all driving forces is the superposition of the different current components, as:

$$I = G\Delta V + G_S \Delta T \qquad (2.21)$$

where the capital *G* here refers to total conductance due to the application of a potential difference, and G_S is the total conductance due to the application of a temperature difference. These are not to be confused with the transport distribution function, where in some cases in the literature is also noted by *G(E)*. In experimental setups, the conductivity due to the application of thermal gradient is extracted by measuring the open circuit voltage V_{OC}. In that case, the open circuit current is by definition zero, and therefore we reach:

$$G_S = -\frac{G V_{OC}}{\Delta T} \Rightarrow -\frac{G_S}{G} = \frac{V_{OC}}{\Delta T} \equiv S \qquad (2.22)$$

where V_{OC} is called the Seebeck voltage, and *S* the Seebeck coefficient defined as:

$$S = -\frac{G_S}{G} = -\frac{I_{(\Delta T \neq 0)}}{\Delta T} \frac{\Delta V}{I_{(\Delta V \neq 0)}} = -\frac{I_{(\Delta T \neq 0)}}{I_{(\Delta V \neq 0)}} \frac{\Delta V}{\Delta T} \tag{2.23}$$

which is just the ratio of the two conductances, the one driven by temperature difference over the one driven by voltage difference. Regarding the maximum power that can be extracted from the thermoelectric generator, this can be determined by a simple circuit analysis as well. For maximum power extraction, the requirement is that the load resistance is equal to the resistance of the thermoelectric material, as $R_L = R_{TE} = 1/G$. In this case, the power delivered to the load (considering that the voltage is split equally in the internal and load resistances) is:

$$P_{\text{max-Load}} = \left(\frac{V_{OC}}{2}\right)^2 / R_L = (S\Delta T)^2 \frac{G}{4} = S^2 G \frac{(T_1 - T_2)^2}{4} \tag{2.24}$$

To derive the efficiency of the thermoelectric engine, we also need to consider the power supplied from the application of temperature gradient in the form of the hear current, J_Q. In a similar manner, we simply assume a heat conductance G_K for the material such that:

$$I_Q = G_K \Delta T \tag{2.25}$$

Finally, the efficiency of the thermoelectric material, the quantity that we are interested in, is the ratio of the maximum possible power that can be extracted from the material (in the form of electrical power), over the power supplied (in the form of heat driven by the temperature gradient) as:

$$\frac{P_{\text{max}}}{P_{\text{supplied}}} = \frac{S^2 \sigma (T_1 - T_2)^2 / 4}{\sigma_K (T_1 - T_2)} = \frac{S^2 \sigma T}{\sigma_K} \frac{(T_1 - T_2)}{4T} = ZT \frac{(T_1 - T_2)}{4T} \tag{2.26}$$

Above $T = \frac{T_1 + T_2}{2}$, and notice that I have use the conductivities rather than conductances, as the difference between the two is just a geometrical factor that gets cancelled from both the numerator and the denominator. The dimensionless quantity ZT is the so-called thermoelectric figure of merit that we seek to maximize, and is defined as:

$$ZT = \frac{S^2 GT}{G_K} \equiv \frac{S^2 \sigma T}{\kappa} \tag{2.27}$$

There is no upper limit for ZT. $ZT \sim 1$ corresponds roughly to 10% of the Carnot efficiency, $\eta_C = (T_H - T_C)/T_H$. The denominator of ZT is the thermal conductivity of the material, adding up heat contributions from phonon transport (the term referred to as the lattice thermal conductivity, κ_l), as well as the heat curried by electrons (the electronic part of the thermal conductivity, κ_e), such that ZT is written commonly as:

$$ZT = \frac{S^2 \sigma T}{\kappa_l + \kappa_e} \tag{2.28}$$

There is a strong activity in the literature towards computing the lattice thermal conductivity, κ_l, of various thermoelectric materials using ab initio methods for the phonon spectrum and the phonon Boltzmann Transport Equation, or Molecular Dynamics and Monte Carlo methods for nanostructures. The methods employed have essentially one-to-one correspondence with the methods that are used to compute electronic conductivity. κ_e, on the other hand, can computed by the electronic BTE as shown below.

To summarize, within the electronic BTE, one can define the moments of the transport distribution function as:

$$R^{(\alpha)} = q_0^2 \int\limits_{E_0}^{\infty} dE \left(-\frac{\partial f_0}{\partial E}\right) \Xi(E) \left(\frac{E - E_F}{k_B T}\right)^{\alpha} \tag{2.29}$$

and from that all relevant transport coefficients as:

$$\sigma = R^{(0)} \tag{2.30a}$$

$$S = \frac{k_B}{q_0} \frac{R^{(1)}}{R^{(0)}} \tag{2.30b}$$

$$\kappa_e = \frac{k_B^2 T}{q_0^2} \left[R^{(2)} - \frac{\left[R^{(1)}\right]^2}{R^{(0)}} \right] \tag{2.30c}$$

To understand the qualitative shape of the conductivity, σ, the Seebeck coefficient, S, and the quantity σS^2 which appears in the numerator of ZT and called the power factor, whose significance we will elaborate on below, at first order we can assume that the relaxation time τ_S is constant and essentially ignore it for qualitative purposes. By assuming that our material under investigation can be described by a single parabolic band, we can numerically compute the trend of the three quantities as a function of the reduced Fermi level, i.e. the distance of the Fermi level from the band edge (as shown in Fig. 2.4), $\eta_F = E_F - E_C$. Clearly, as the Fermi level is placed nearer and eventually higher into the bands, the carrier density increases exponentially, which makes the conductivity to exponentially increase as well. On the other hand, a careful examination of (2.30b) reveals that at first order, the Seebeck coefficient is linear (with a negative slope) in η_F as the conductivity terms appear in the denominator and numerator, and at first order 'cancel out'. Thus, as the Fermi level is raised, and η_F is increased, the Seebeck coefficient drops. In the non-degenerate limit (i.e. when E_F is still in the gap, before reaching close to the bands), the Seebeck coefficient experiences a linear drop, something easily observed when the Fermi statistics are approximated by Boltzmann statistics in (2.30b). In the degenerate

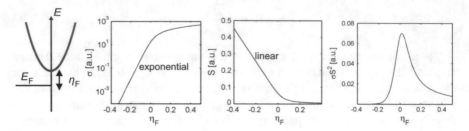

Fig. 2.4 The quantitative shape of the thermoelectric coefficients conductivity, σ, Seebeck coefficient, S, and the quantity σS^2 versus the reduced Fermi level, defined as the distance of the Fermi level from the band edge $\eta_F = E_F - E_C$

limit, this approximation is not valid, and the Fermi-Dirac statistics make the trend more complex, such that the Seebeck coefficient experiences a less dramatic drop. Note, however, that the Seebeck coefficient of n-type materials, as the one in Fig. 2.4, is negative. It is only customary to plot the absolute values of the Seebeck coefficient. Finally, the power factor σS^2, which determines the numerator of the ZT figure of merit is plotted in Fig. 2.4 as well, indicating a maximum, typically around $\eta_F = 0$.

There is a connection between the electronic part of the thermal conductivity (κ_e) and the electronic conductivity σ through the Wiedemann-Franz law, i.e. $\kappa_e = L\sigma T$. The proportionality constant L is called the Lorenz number (not to be confused with length). For metals and degenerate semiconductors, L reaches the Sommerfeld value, $L_0 = \pi^2/3(k_B/q)^2 = 2.45 \times 10^{-8}$ W Ω K^{-2} where k_B and q are the Boltzmann constant and charge of the electron, respectively. This value drops to $L_0 = 2(k_B/q)^2 = 1.49 \times 10^{-8}$ W Ω K^{-2} for non-degenerate, single parabolic band materials and acoustic phonon scattering conditions. The Lorenz number plays an important role in the experimental determination of the phonon, or lattice part of the thermal conductivity (κ_L) from thermal conductivity measurements. This is done by using the Wiedemann-Franz law to estimate the electronic part of the thermal conductivity (κ_e) from electronic conductivity measurements (σ). κ_e is then subtracted from the experimentally measured value of the total thermal conductivity (κ_{tot}). The Lorenz number plays an important role in thermoelectric materials, and essentially ways to reduce it are explored, which often provide improved ZT values, as it usually corresponds to reduction in the electronic part of the thermal conductivity. Important deviations in the Wiedemann-Franz law are encountered in several cases: multi-band materials [12, 13], confined dimensions [14] such as nanowires [15], quasi-1D systems [16] and effective 0-D systems such as quantum dots [17], single molecule [18] and single atom systems [19]; under conditions of quantum criticality [20], in superconductors [21], in superlattices and granular metals [22], and in the presence of disorder [23]. All the above categories are active areas of research.

In the case of bipolar transport in small bandgap materials, in which case as the temperature is raised both majority and minority carriers are excited and participate in transport, an additional thermal conductivity component arises and is added to the corresponding majority and minority parts. That is called the bipolar thermal conductivity, κ_{bi}, given by:

$$\kappa_{bi} = \frac{\sigma_e \sigma_h}{\sigma_e + \sigma_h}(S_e - S_h)^2 T \tag{2.31}$$

which makes the total electronic part of the thermal conductivity a summation of three quantities, $\kappa_{tot} = \kappa_{el} + \kappa_h + \kappa_{bi}$. This becomes particularly important for example in PbTe at temperatures above 700 K, and needs to be accounted properly in calculations [24]. As the Seebeck coefficients of holes and electrons have opposite signs, the bipolar thermal conductivity is highest at midgap, when both Seebeck values are high. The different signs of S_e and S_h, on the other hand, reduce the overall Seebeck coefficient of the material when E_F is placed at midgap. Thus, a bipolar material suffers from larger thermal conductivity, but also reduced power factor through cancellations of components of the Seebeck coefficients of electrons and holes. It should also be noted that κ_{bi} is a conductivity-limited quantity, and is primarily determined by the conductivity of the minority carrier (the smaller of the two as seen by the left term in (2.31)). This can also be seen from its physical origin of electron-hole recombination in the contacts [25]. Efforts to reduce bipolar conduction include the presence of nanoinclusions and grains with potential barriers for minority carrier transport, methods which are most effective when the minority carrier mean-free-paths are longer compared to the majority carrier mean-free-paths [26].

2.4 Carrier Scattering

Despite the relative 'straight-forwardness' in obtaining the density of states and bandstructure velocities that are needed to extract the TE coefficients through DFT first principle calculations (for which a large number of software packages are available [27–30]), the momentum relaxations times are much less straight-forward to extract. In most TE studies in the literature, when it comes to complex bandstructure materials it is common to employ a constant relaxation time, usually of value $\tau_S = 10^{-14}$ s [2]. This is because of the huge complexity in extracting energy dependent values for arbitrary bandstructures. Recently, more sophisticated calculations using the Electron-Phonon Wannier (EPW) simulator package [31], allow the extraction of the electron-phonon scattering rates from first principles. However, these simulations are computationally extremely expensive requiring dense electron and phonon dispersions, and only treat electron-phonon scattering, whereas at the densities in which the power factor peaks, it is ionized impurity scattering that is the dominant scattering mechanism [9, 10, 32]. A middle ground, is to extract energy dependent scattering rates using some parameters, such as deformation potentials, the dominant phonon energies, and dielectric constants. In general, the use of energy

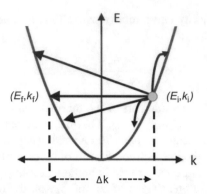

Fig. 2.5 A schematic of the scattering processes that a carrier in an initial state (E_i, k_i) in a parabolic band undergoes under elastic (energy invariant) and inelastic (energy changing) scattering processes. Isotropic scattering processes do not depend on the momentum exchange Δk, but anisotropic processes do so. The largest Δk is, the weaker the scattering process is

dependent scattering methods has been the norm in the semiconductor electronic transport community [3, 4].

The formal derivation of scattering from the quantum mechanical level is beyond the goal of this brief, thus, I only provide a brief description of the scattering mechanisms, and then describe the numerical implementation of their use for transport in complex bandstructure materials. In the event of scattering, a carrier residing in an initial state with a given energy, momentum and band *(E, k, n)*, scatters after the interaction with a perturbation U, to a final state *(E', k', m)* as pictured in Fig. 2.5.

The carrier in the initial state has velocity v_i, given by the slope of the band at that k-point. After scattering that velocity can change. If the carrier scatters to a state with negative velocity we refer to it as back-scattering, but it can also scatter in states with positive velocities as well, thus, it is important that the momentum relaxation time (or velocity relaxation time more accurately) is employed in calculations, as not all scattering events will randomize their initial motion the same.

The scattering description begins with the formulation of the quantum mechanical scattering matrix element, following Fermi's Golden Rule, which employs the wavefunctions of the initial and final states together with the perturbing potential as [3, 32]:

$$H_{k',k}^{m,n} = \frac{1}{\Omega} \int_R \psi_m(\vec{r})^* U_S(\vec{r}) \psi_n(\vec{r}) \mathrm{d}r \tag{2.32}$$

Scattering can be elastic, if the energy of the initial and final states is the same (such as for acoustic phonon scattering and ionized impurity scattering), or inelastic, if the energy of the initial and final states differs (as for optical phonon scattering events). From the matrix element we then define the transition rate as the square of the matrix element for all states that obey energy conservation according to the scattering process considered, as:

$$S_{n,m}(k, k') = \frac{2\pi}{\hbar}|H_{k',k}^{m,n}|^2 \delta(E_m(k') - E_n(k) - \Delta E)$$ (2.33)

Finally, the momentum relaxation scattering rate is defined by inserting an extra term that accounts for the difference in the momentum of the initial and final states as:

$$\frac{1}{\tau_n(k)} = \sum_{m,k'} S_{n,m}(k, k')\left(1 - \frac{|p_m(k')|}{|p_n(k)|}\cos\vartheta\right)$$ (2.34)

In a parabolic band case, the momentum of a carrier is easily defined from the bandstructure as $p = \hbar k$, and every k-state has a well-defined velocity. In the case of complex bandstructure materials as is the case of most thermoelectric materials, however, when bands appear in different parts of the Brillouin zone, it is convenient to use the velocity of carriers (in the direction of transport, say in x-), rather than the momentum, as:

$$\frac{1}{\tau_n(k_x)} = \sum_{m,k_x'} S_{n,m}(k, k')\left(1 - \frac{v_m(k_x')}{v_n(k_x)}\right)$$ (2.35)

The main scattering mechanisms we consider in the case of thermoelectric materials are phonon scattering (acoustic and optical including polar optical), ionized impurity scattering, alloy scattering, and boundary scattering. Below, we present the scattering rate equations without going into the derivations, as those can be found all over the literature [3].

The scattering rate for acoustic phonon scattering is elastic and isotropic (does not depend on the momentum difference between the initial and final states) and is given by:

$$\frac{1}{\tau_{ADP}(E)} = \frac{\pi D_A^2 k_B T}{\hbar \rho v_s^2} g(E)$$ (2.36)

where D_A is the deformation potential (representing the scattering strength), ρ is the mass density of the material, and v_s is the sound velocity.

The scattering rate for optical phonon scattering is given by:

$$\frac{1}{\tau_{ODP}(E)} = \frac{\pi D_O^2}{2\rho\omega_{ph}}\left(N_\omega + \frac{1}{2} \mp \frac{1}{2}\right)g(E \pm \hbar\omega_{ph})$$ (2.37)

where D_O is the optical deformation potential, ω_{ph} is the optical phonon frequency, and N_ω is the Bose–Einstein distribution function given by:

$$N_\omega = \frac{1}{\exp(\hbar\omega_{ph}/k_B T) - 1} \tag{2.38}$$

For ionized impurity scattering, the most common model used is the Brooks-Herring model, under which the scattering rates are given by:

$$\tau_{IIS}(E) = \frac{16\sqrt{2m^*}\pi k_s^2 \varepsilon_0^2}{N_I q^4}\left[\ln(1+\gamma^2) - \frac{\gamma^2}{(1+\gamma^2)}\right]E^{3/2}$$

$$\gamma^2 = \frac{8m_{DOS}(E-E_0)L_D^2}{\hbar^2}, \quad L_D = \sqrt{\frac{\kappa_s\varepsilon_0}{q}\frac{\partial E_F}{\partial n}} \approx \sqrt{\frac{\kappa_s\varepsilon_0 k_B T}{q^2 n}\frac{\Im_{1/2}(\tilde{\eta}_F)}{\Im_{-1/2}(\tilde{\eta}_F)}} \tag{2.39}$$

where L_D is the screening length, which depends on the dielectric constant, κ_S and the density n and its derivative with the Fermi level E_F, which is in general temperature and doping dependent. The explicit use of $\partial n/\partial E_F$ enables us to apply the equation also in the degenerate doping conditions. For simple bands equivalently one can use the second expression, which includes the Fermi Dirac distributions, but for generic bands, the more accurate expression is the first one, which takes directly the variation of the charge with the Fermi level. We note, however, that the discrepancies between the two methods are minimal. N_I is the impurity density (the doping density), and n is the carrier density, which can differ from N_I, for example in depletion regions, or spatially under modulation doping cases. This is an important scattering mechanism as TE materials are operating under highly doped, degenerate conditions. The ionized impurity scattering rates are elastic and anisotropic, since a scattering event does not change the carrier's energy, but changes is momentum with the scattering rate depending on the length of the momentum (wavevector) exchange vector $\Delta k = k - k'$. The scattering rate becomes weaker at higher energies when the exchange momentum vector is large (i.e. it favours narrow angles), and depends linearly on the number of impurities. At very high impurity densities, the ionized impurities are strongly screened by the electron cloud, and the scattering rate converges to an isotropic relation, proportional to the density of states as in the case of electron-phonon scattering, and the impurity number, as:

$$\frac{1}{\tau_{IIS}(E)} = \frac{\pi N_I}{\hbar}\left(\frac{qL_D}{k_s\varepsilon_0}\right)^2 g(E) \tag{2.40}$$

In all cases, scattering is proportional to the density of final states that the carriers can scatter into. For example, the scattering rate for a simple parabolic band of $m^* = 1m_0$ and acoustic phonon scattering with $D_A = 5$ eV is shown in Fig. 2.6a, following the $E^{1/2}$ trend as for the density-of-states in 3D. On the other hand, the ionized impurity scattering has a different trend as in Fig. 2.6b depending on the carrier density. At lower densities, the low-energy carriers scatter more strongly, and as the density increases, the rates begin to follow the density-of-states trend as well.

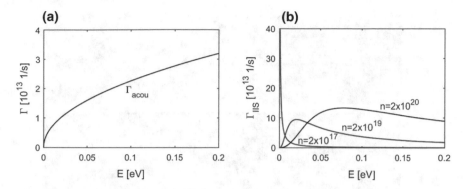

Fig. 2.6 **a** The scattering rate versus energy for acoustic phonon scattering. **b** The scattering rate versus energy for ionised impurity scattering at three different dopant densities: $n = 2 \times 10^{17}$ cm^{-3} (blue line), $n = 2 \times 10^{19}$ cm^{-3} (red line), $n = 2 \times 10^{20}$ cm^{-3} (black line). A simple parabolic band of $m^* = 1m_0$ and $D_A = 5$ eV are assumed [33]

Another important scattering mechanism for TE materials is alloy scattering, since large families of materials are indeed alloys. Alloy scattering has its origin in the random variation of the crystal potential that the carriers encounter as they travel through the lattice. The scattering rate, (as derived for crystals with zincblende structure), is given by [3, 34, 35]:

$$\frac{1}{\tau_{\text{alloy}}} = \frac{3\pi^3}{16\hbar} x(1-x)\Omega(V_A - V_B)^2 g(E) \tag{2.41}$$

where V_A and V_B are the potential of the lattice that each element introduces, Ω is the volume of the primitive cell and $g(E)$ is the density of states of the alloy, as obtained by interpolating the density of states of the constituent compounds.

Finally, in the case of nanostructured thermoelectric materials, boundary scattering needs to be accounted for as well. Boundary scattering is in general a complex mechanism, and cannot be easily treated accurately. We discuss real space methods in the following chapters that could account for boundary scattering. Depending on the nature of boundaries, whether they are specular or randomizing, depending if they are localized (as in the case of nanoinclusions) or elongated (as in grain boundaries and superlattices), depending if they introduce large or small potential barriers, the scattering rate is different, and thus, it becomes very difficult to extract and model it accurately. Examples are depicted in Fig. 2.7. An effective way to model the rate of this scattering event is to define it as the velocity of electrons divided by an effective distance at which the carriers undergo scattering. Simply, a mean-free-path for scattering is the effective distance between defects (d_{eff}). A constant C can be usually included to describe the strength of the scattering event and match experimental data as:

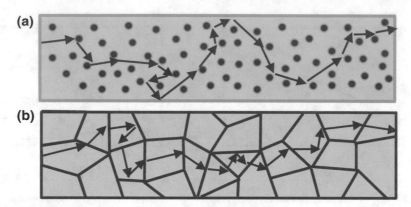

Fig. 2.7 **a** Schematics of an electron trajectory through a matrix material in the presence of nanoinclusion scattering centers. **b** Schematic of an electron trajectory through a nano/poly-crystalline material composed of grains and grain boundaries

$$\frac{1}{\tau_{\text{defect/boundary}}} = C\left(\frac{\upsilon(E)}{d_{\text{eff}}}\right) \tag{2.42}$$

In the case of grain boundaries or interfaces between dissimilar materials, in which case potential barriers of average height V_B are formed for carriers, it is customary to impose thermionic emission above the barriers as:

$$\sigma(E) = 0 \text{ for } E <= V_B \tag{2.43a}$$

$$\sigma(E) = \sigma_0(E)T_r(E) \text{ for } E > V_B \tag{2.43b}$$

In this case, transport over the barrier provides energy filtering, which improves the Seebeck coefficient, at the expense of the conductivity. The transmission, T_r, used, is usually a step function, which takes zero values below V_B, and unity above V_B. However, this can be quite generic, and the transmission can be a function that includes tunnelling, or quantum reflections, etc., something that can be extracted from typical quantum mechanical simulations, or methods like the WKB approximation. In particular, although a step function transmission is the easier to be used, it does not account for well/barrier momentum state mismatch/conservation, and typically overestimates the conductivity over the barrier.

Finally, the strength of all scattering mechanisms is combined to an overall strength using Matthiessen's rule, under which the total scattering time is then given by [36]:

$$\frac{1}{\tau_{\text{tot}}} = \frac{1}{\tau_{\text{ADP}}} + \frac{1}{\tau_{\text{ODP}}} + \frac{1}{\tau_{\text{IIS}}} + \frac{1}{\tau_{\text{alloy}}} + \frac{1}{\tau_B} + \dots \tag{2.44}$$

In the equation above, the rate is computed at each energy E, for each k-state, but for brevity the indices are not inserted in the equation.

2.5 Ballistic Transport—The Landauer Approach

There are cases where given a certain bandstructure $E(k)$ for a TE material, one can assess its ballistic transport TE performance, i.e. in the absence of scattering processes. Although this scenario is rarely realistic, such studies are useful as they can provide qualitative information about the isolated influence of the carrier velocities or the numbers of modes/bands that the material has on the TE coefficients, and especially on the Seebeck coefficient. If we count the contribution of all k-space states in the dispersion, and extract their velocity from the gradient of the bandstructure, the current can be evaluated as the product of charge times velocity. In k-space the number of states in a discretized dk^d space (where d is the dimensionality of the system), is well defined, constant, and depends on the length of the Brillouin zone and the domain discretization. Having the velocities of all states, the conductance G, the Seebeck coefficient S, the PF, and the electronic part of the thermal conductivity κ_e can be extracted as indicated above in (2.30a)–(2.30c) by using the auxiliary functions as [37]:

$$R^{(\alpha)} = \frac{-\frac{q^2}{L} \sum_{k_\parallel > 0} v_k (f_1 - f_2)(E_k - \mu_1)^\alpha}{(\mu_1 - \mu_2)}, \tag{2.45}$$

where v_k is the bandstructure velocity, and f_1, f_2 are the Fermi functions of the left and right contacts, respectively, μ_1, μ_2 are the contact chemical potentials (for the purposes of this discussion they are equivalent to the Fermi levels of the contacts—expressed in Volts), and E_k is the subband dispersion relation. For small driving fields ΔV, the linearization $f_1 - f_2 = -q_0 \Delta V \frac{\partial f_1}{\partial E}$ holds, where $\Delta V = \mu_1 - \mu_2$, and this relates back to the linear response in BTE. Here, however, the computation is explicitly performed in k-space rather than energy-space. Note that because in the absence of the relaxation time in (2.45) (as we have considered only ballistic conditions), we obtain the conductance G, instead of conductivity σ. To relate these results to experiments, one has to consider the materials' geometry (multiply by the cross sectional area and divide by the materials' length), and it is common to multiply (2.45) by a mean-free-path for scattering, λ, which can map the conductivity into the conductance (also with proper units) [38]. In most studies, however, ballistic results are used for comparative studies, rather than extracting absolute values, such that real measurement complications are avoided. The mean-free-path can be expressed as $\lambda = v\tau$, which at the end makes (2.45) to map to (2.29) and (2.30a), and transforms G into σ. However, although this treatment simplifies things a lot, it still requires knowledge of λ (or even the more detailed k-resolved λ_k), which is not always available information.

2.6   Numerical Extraction of DOS (E), $\nu(E)$, and $\tau(E)$ from Arbitrary Bandstructures

The most common way to extract the thermoelectric coefficients of complex band-structure TE materials, is to compute in some way the relaxation times, the carrier velocities, and their density of states, as a function of energy. The density and velocity of each state are quantities determined directly from the bandstructure of materials.

Typical electronic structure codes such as DFT and tight-binding, compute the $E(k)$ on a regular k-discretized grid in the three Cartesian coordinates. If the scattering times for each k-state were also known, then a simple numerical summation over the Brillouin zone will have sufficed in obtaining the transport properties. The equations that describe the scattering rates, however, are functions of the density-of-states in energy. The availability of the state attributes in a regular k-grid, rather than a regular E-grid that is required for the $\tau(E)$, imposes a significant computational issue in these calculations, as the uniform k-grid resolved $E(k)$ needs to be converted into a regular E-grid, referred to as '$k(E)$'. In essence, the iso-energy surfaces (in 3D) or lines (in 2D), that form the electronic structure, and from there the density-of-states at all energies, need to be evaluated. In practice, however, in a fully numerical calculation that we are interested in, we need to group all states according to their energy and do so for each band (or valley) as well. In general, all k-states will have numerically different energies (unless they obey crystal symmetry rules, but we leave this for now)—maybe similar in some cases, but different in the strict mathematical sense. Thus, in order to form the iso-energy surfaces from groups of k-space points, interpolation schemes need to be employed. Once the iso-energy surfaces (3D) or lines (2D) are formed, then the area/length of the surface/contour, together with the perpendicular bandstructure velocities as I will explain below, allows for the extraction of the DOS(E).

I will start with discussing the process of extracting the relevant quantities in 2D, which is conceptually and numerically simpler, and then describe the 3D process. In 2D the procedure of forming iso-energy contours is depicted in Fig. 2.8. For illustration regarding the contour extraction, I consider a circular contour in the $k_x - k_y$ space as shown in Fig. 2.8 (i.e. dashed-green line), although the approach works for arbitrary contour shapes.

To construct the contour of a subband at a specific energy, the following steps are performed: (i) We divide the rectangular k-space grid (Fig. 2.8a) into triangles as shown in Fig. 2.8b and keep record of the coordinates and the energies E_A, E_B, and E_C of the three triangle corners as shown in Fig. 2.8c. (ii) We find the eigen-energies of the specific subband into consideration that reside around the energy of interest (within a lower and upper energy cutoffs) and record the triangles which have the energy of one of their corners above/below and the energies of the other two corners below/above that energy level. This indicates that the energy contour passes through that triangle as shown in Fig. 2.8c. (iii) The third step is to compute the k-space length of the contour segments in each of the relevant triangles. This is done by simple geometric considerations. The $k_{x, y}$-points on the triangle sides at energy E

2.6 Numerical Extraction of DOS (E), $v(E)$, and $\tau(E)$ ...

29

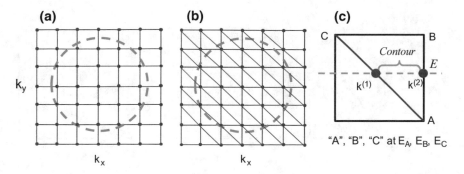

Fig. 2.8 The process in which an iso-energy contour is extracted from a rectangular k-grid. **a** The rectangular k-grid and the iso-energy contour. **b** The triangulation of the rectangular grid. **c** The contour element length between k-points on the rectangular and triangular segments [39]

(the red labeled dots $k^{(1)}$, $k^{(2)}$ in Fig. 2.7c) are given by:

$$k_{x/y}^{(1/2)} = k_{x/y}^A + \left(k_{x/y}^{B/C} - k_{x/y}^A\right)\frac{\left(E - E^A\right)}{\left(E_{x/y}^{B/C} - E_{x/y}^A\right)} \tag{2.46}$$

The length of the individual segments is then computed as:

$$dL_{k,n,E}^{(1\to2)} = \sqrt{\left(k_x^{(1)} - k_x^{(2)}\right)^2 + \left(k_y^{(1)} - k_y^{(2)}\right)^2} \tag{2.47}$$

Once the individual segment lengths dk are extracted for a specific energy at a given subband, the energy resolved density of states of the particular element on the contour is computed by $g_{k,n,E}^{2D} = \frac{dL_{k,n,E}}{\hbar\langle v_n(k_x,k_y,E)\rangle}$, where $dL_{k,n,E}$ is the generalized dk element length, and $\langle v_n(k_x, k_y, E)\rangle$ is the average group velocity of the states in subband n at energy E along the contour C_k computed as $v_n(k_x, k_y, E) = \sqrt{v_n(K_x, E)^2 + v_n(k_y, E)^2}$. Each component of the directional velocities is given by the gradient of the bandstructure as $v_n(k_{x/y}, E) = \frac{1}{\hbar}\partial E_n(k_{x/y})/\partial k_{x/y}$, where $E_n(k_{x/y})$ is the energy of the particular state in subband n. For better accuracy, for the velocity $\langle v_n(k_x, k_y, E)\rangle$ we use the average of the velocities of the bandstructure at the three points of their corresponding triangle. Then the overall density of states at that energy is simply the integration of $g_{k,n,E}^{2D}$ along the contour C_k, or numerically it is the summation of the DOS of all individual segments as:

$$g_n^{2D}(E) = \left(\frac{2}{4\pi^2}\right)\sum_{C_k}\frac{dL_{k,n,E}}{\hbar\langle v_n(k_x, k_y, E)\rangle}, \tag{2.48}$$

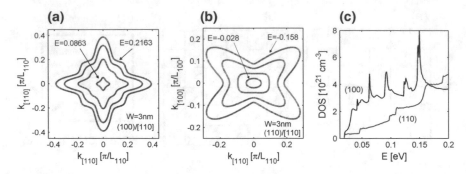

Fig. 2.9 Constant energy contours for the valence band of Si ultra-thin-layers of width $W = 3$ nm, in the (100) and (110) confinement surfaces. In **c** the corresponding density of states are shown as a function of energy [39]

where we have included the 2 for the spin, and the factor $1/(2\pi)^2$ in the denominator comes from the conversion of the k-space summation into integration in energy (of course numerically we evaluate a summation in energy as well).

Beyond parabolic/non-parabolic bands which have circular energy contours, this method is generic enough to be applied to warped bands as well. Figure 2.9a, b, for example, show typical energy contours in p-type Si 2D ultra-thin-layers (UTLs) of thicknesses $W = 3$ nm on the (100) and (110) surfaces, respectively. Contours at various different energies are shown, indicating how the warped bandstructure features and anisotropy are captured. The energy resolved density of states DOS(E) for the two surfaces is shown in Fig. 2.9c. The deviations from the flat plateaus in the DOS(E) expected for a parabolic 2D subband, point to the non-parabolicity of the bands, whereas the discontinuities point to the onsets of the different subbands.

In a similar manner, the extension to 3D can be formed, which is the relevant case for bulk, complex TE materials. The 3D energy surface can be in general built by using the tetrahedron approach [40] by means of a local Delaunay Triangulation (DT), where the reciprocal unit cell is sampled by a regular mesh of 3D elements. All mesh elements are scanned numerically to check if the 3D constant energy surface passes through them. For this the energy of all the vertex pairs (12 pairs in 3D) are checked, for the condition if at least one of these pairs consists of a vertex with energy above the value of interest (the energy surface value under consideration) and one below. If so, it means that the specific constant energy contour passes through the element under consideration. All the relevant elements are then triangulated, forming tedrahendra, whose vertices will be 'cut' by the energy surface. The $E(k)$ is then interpolated linearly between these points as earlier in the 2D description to obtain the estimated k-point at the energy of interest, and the surface area of the triangles formed, dA_k, is computed using geometrical considerations—that area, together with the perpendicular velocity vector will determine the density-of-states associated with that element of the iso-energy surface as $g_{k,n,E} = \frac{dA_{k,n,E}}{\hbar|\overrightarrow{v_{k,n,E}}|}$.

As simpler, and computationally faster approach can be followed by avoiding the triangulation of the reciprocal unit cell, as described in [10]. In this case, a search through all k-points' energies and their nearest neighbors is performed, to identify if the iso-energy of interest passes through them. Then a linear interpolation is performed along the edges connecting the nearest neighbors. A collection of k-points that belong to the iso-energy of interest is then formed. The difference from the Delaunay triangulation method is that the energy surface itself is not constructed. Because of that, a surface element area dA_k then needs to be assigned for each point in a different way. For this, for every relevant k-point, the nearest neighbours are explored in a radius of length $\sqrt{2}dk$, where dk is the distance between adjacent points in the initial mesh, the discretization length used to calculate the bands to begin with. Then, we calculate the average distance between the given point and its detected neighbours, $<\Delta k>$. The surface element associated to the k-point is approximated by a circle of radius half the average distance to the neighbouring points, i.e. $dA_k = \pi\left(\frac{<\Delta k>}{2}\right)^2$. The approximation provides very good agreement in evaluating the DOS from simple parabolic band cases, to complex bandstructure materials such as half-Heuslers [9, 10].

Therefore, for a fully numerical scheme, we go back to the TDF as in (2.17), which is now defined by individually accounting the contribution to transport from all states as:

$$\Xi(E) = \sum_{k,n}^{BZ} v_{k,n,E}^2 \tau_{k,n,E} \delta(E_k - E) = \frac{2}{(2\pi)^3} \sum_{k,n}^{\mathcal{L}_E^n} v_{k,n,E}^2 \tau_{k,n,E} g_{k,n,E} \qquad (2.49)$$

where $v_{k,n,E}$ is the band velocity of the transport state defined by the wave vector k in band n at energy E, $\tau_{k,n,E}$ is its momentum relaxation time (combining the relaxation times of each scattering mechanism using Matthiessen's rule), and $g_{k,n,E}$ its density-of-states (DOS). BZ stands for the Brillouin Zone and $\mathcal{L}_E^n$ represents the surface of constant energy E for the band of index n. The first sum in Eq. (2.49) runs over all the k-states and the bands of the BZ. The delta-function which picks up only the states at energy E, defines a surface of constant energy E for each band. The second sum in Eq. (2.49) runs over all the points on these surfaces, for all the bands, and returns an energy dependent quantity for each band, evaluated on the iso-energy surface at E. The triad k, n, E defines uniquely each transport state [10].

Each scattering relaxation time above is defined by the summation of all transition rates of an initial state (E, k, n) to all available final scattering states (E', k', m) for each scattering mechanism. For acoustic phonons this is:

$$\frac{1}{\tau_{ADP}(E)} = \frac{\pi D_A^2 k_B T}{\hbar \rho v_s^2} \sum_{m,k'} g_{k'}(E)\left(1 - \frac{v_m(k_x')}{v_n(k_x)}\right) \qquad (2.50)$$

For optical phonons:

$$\frac{1}{\tau_{ph}^{ODP}(E)} = \frac{\pi}{\hbar} \frac{\left(N_\omega + \frac{1}{2} \mp \frac{1}{2}\right)}{\rho \hbar \omega_{ph}} \sum_{m,k'} g_{k'}\left(E \pm \hbar \omega_{ph}\right)\left(1 - \frac{v_m(k_x')}{v_n(k_x)}\right) \qquad (2.51)$$

For polar optical phonons:

$$\frac{1}{\tau_{POP}(E)} = \frac{\pi q^2 \omega_{ph}}{\varepsilon_0}\left(\frac{1}{\kappa_\infty} - \frac{1}{\kappa_s}\right)\left(N_\omega + \frac{1}{2} \mp \frac{1}{2}\right)\sum_{m,k'}\frac{g_{k'}\left(E \pm \hbar\omega_{ph}\right)}{|k - k'|^2}\left(1 - \frac{v_m\left(k'_x\right)}{v_n(k_x)}\right)$$

(2.52)

For ionized impurity scattering:

$$\frac{1}{\tau_{IIS}(E)} = \frac{2\pi}{\hbar}\frac{Z^2 q^4 N_{imp}}{\kappa_s^2 \varepsilon_0^2}\sum_{k'_x}\frac{g_{k'}\left(E \pm \hbar\omega_{ph}\right)}{\left(|k - k'|^2 + \frac{1}{L_D^2}\right)^2}\left(1 - \frac{v_m\left(k'_x\right)}{v_n(k_x)}\right)$$

(2.53)

These equations sum up all different states for every iso-energy surface and for all relevant scattering mechanisms, to extract the overall scattering rate and then the contribution of all states to the TDF that is needed to extract the conductivity

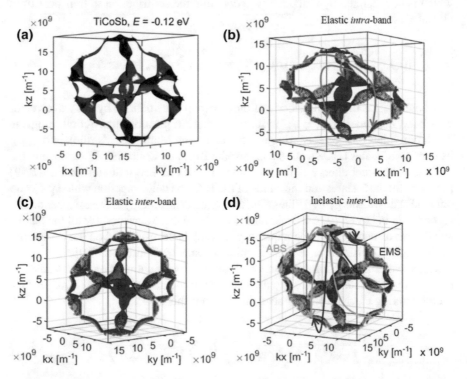

Fig. 2.10 **a** Constant energy surface for the valence band of the half-Heusler TiCoSb at energy $E = -0.12$ eV into the valence band. **b** The energy surface k-states with the arrows indicate the transition processes for different elastic intra-band scattering events. **c** Red dots show the available states in another band for elastic inter-band transitions. **d** The inelastic inter-band transitions due to the emission/absorption of optical phonons [10]

and the rest of the TE coefficients. An example of a complex bandstructure material is indicated above for the TiCoSb half-Heusler in Fig. 2.10. The complex energy surfaces are for an energy $E = -0.12$ eV into the valence band. The iso-energy surface has a large number of valleys, which are beneficial for TEs. It also has elongated features which indicate large anisotropy in the effective mass, another beneficial property for TE materials, pointing out to potential optimization routes for high power factors. The different arrows also indicate the various scattering events that take place for a few examples of transitions from initial to final states in the same iso-surface (**blue state to blue state**), or inelastic processes on other energy surfaces (**blue states to orange/green** states).

Note that with respect to phonon scattering, what is described here is based on the deformation potential theory. In principle, different values can be used for different scattering events and different band selections rules (as for example for the g- and f-processes in Si [41]). In a more fundamental sense, the actual transition rates can be computed by taking into account the phonon spectrum, phonon wavefunctions, and electronic wavefunctions in addition, which provides a fully numerical evaluation that captures the various selection rules as well. These are computationally much more expensive routes, but codes have been developed to provide such capabilities, such as the EPW software package [31]. On the other hand, such codes need to be used with care as well, since they compute the scattering relaxation times, but not the momentum (or velocity) relaxation times, which are used within the BTE. In the case of ADP or ODP scattering these provide the same TDFs due to symmetry. This is not true for polar optical phonons (POP) or for ionised impurity scattering however (for which rates are not provided within the EPW software anyway).

2.7 Conclusions

This Chapter introduced the basics of the Boltzmann Transport formalism that is commonly employed in the extraction of thermoelectric coefficients and elaborated how the actual coefficients and figure-of-merit are extracted from simple circuit theory considerations. Then it discussed the different scattering mechanisms that contribute to the overall relaxation scattering times used in the BTE formalism. Finally, it described a numerical scheme that allows the description of the relaxation times as function of energy, and their calculation for each state independently, such that they can be applied for bandstructures of arbitrary shapes. Such energy dependent scattering rates provide different energy dependences for the TDF and different trends for the TE coefficients compared to constant relaxation time considerations, and could lead to better predictions and optimization routes.

References

1. Madsen, G.K., Singh, D.J.: BoltzTraP. A code for calculating band-structure dependent quantities. Comput. Phys. Commun. **175**(1), 67–71 (2006)
2. Pizzi, G., Volja, D., Kozinsky, B., Fornari, M., Marzari, N.: BoltzWann: a code for the evaluation of thermoelectric and electronic transport properties with a maximally-localized Wannier functions basis. Comput. Phys. Commun. **185**(1), 422–429 (2014)
3. Lundstrom, M.: Fundamentals of Carrier Transport. Cambridge University Press, Cambridge (2000)
4. Nag, B.R.: Electron Transport in Compound Semiconductors. Springer Series in Solid-State Sciences (SSSOL), vol. 11. Springer, Berlin (1980)
5. Ziman, J.M.: Electrons and Phonons: The Theory of Transport Phenomena in Solids. Oxford Classic Texts in the Physical Sciences. Oxford University Press, Oxford (2001)
6. Pierret, R.F.: Semiconductor Device Fundamentals. Addison-Wesley, Boston (1996)
7. Mahan, G.D., Sofo, J.O.: The best thermoelectric. Proc. Natl. Acad. Sci. **93**(15), 7436–7439 (1996)
8. Neophytou, N., Kosina, H.: Atomistic simulations of low-field mobility in Si nanowires: influence of confinement and orientation. Phys. Rev. B **84**(8), 085313 (2011)
9. Kumarasinghe, C., Neophytou, N.: Band alignment and scattering considerations for enhancing the thermoelectric power factor of complex materials: The case of Co-based half-Heusler alloys. Phys. Rev. B **99**(19), 195202 (2019)
10. Graziosi, Patrizio, Kumarasinghe, Chathurangi, Neophytou, Neophytos: Impact of the scattering physics on the power factor of complex thermoelectric materials. J. Appl. Phys. **126**, 055105 (2019)
11. Datta, S.: Lessons from Nanoelectronics: A New Perspective on Transport, vol. 1. World Scientific Publishing Company, Singapore (2012)
12. Thesberg, M., Kosina, H., Neophytou, N.: On the Lorenz number of multiband materials. Phys. Rev. B **95**(12), 125206 (2017)
13. Zhao, L.D., Lo, S.H., Zhang, Y., Sun, H., Tan, G., Uher, C., Wolverton, C., Dravid, V.P., Kanatzidis, M.G.: Ultralow thermal conductivity and high thermoelectric figure of merit in SnSe crystals. Nature **508**(7496), 373 (2014)
14. Ou, M.N., Yang, T.J., Harutyunyan, S.R., Chen, Y.Y., Chen, C.D., Lai, S.J.: Electrical and thermal transport in single nickel nanowire. Appl. Phys. Lett. **92**(6), 063101 (2008)
15. Völklein, F., Reith, H., Cornelius, T.W., Rauber, M., Neumann, R.: The experimental investigation of thermal conductivity and the Wiedemann-Franz law for single metallic nanowires. Nanotechnology **20**(32), 325706 (2009)
16. Casian, A.: Violation of the Wiedemann-Franz law in quasi-one-dimensional organic crystals. Phys. Rev. B **81**(15), 155415 (2010)
17. López, R., Sánchez, D.: Nonlinear heat transport in mesoscopic conductors: rectification, Peltier effect, and Wiedemann-Franz law. Phys. Rev. B **88**(4), 045129 (2013)
18. Wang, R.Q., Sheng, L., Shen, R., Wang, B., Xing, D.Y.: Thermoelectric effect in single-molecule-magnet junctions. Phys. Rev. Lett. **105**(5), 057202 (2010)
19. Kubala, B., König, J., Pekola, J.: Violation of the Wiedemann-Franz law in a single-electron transistor. Phys. Rev. Lett. **100**(6), 066801 (2008)
20. Tanatar, M.A., Paglione, J., Petrovic, C., Taillefer, L.: Anisotropic violation of the Wiedemann-Franz law at a quantum critical point. Science **316**(5829), 1320–1322 (2007)
21. Graf, M.J., Yip, S.K., Sauls, J.A., Rainer, D.: Electronic thermal conductivity and the Wiedemann-Franz law for unconventional superconductors. Phys. Rev. B **53**(22), 15147 (1996)
22. Tripathi, V., Loh, Y.L.: Thermal conductivity of a granular metal. Phys. Rev. Lett. **96**(4), 046805 (2006)
23. Sun, X.F., Lin, B., Zhao, X., Li, L., Komiya, S., Tsukada, I., Ando, Y.: Deviation from the Wiedemann-Franz law induced by nonmagnetic impurities in overdoped La 2–x Sr x CuO 4. Phys. Rev. B **80**(10), 104510 (2009)

24. Kim, H.S., Lee, K.H., Yoo, J., Shin, W.H., Roh, J.W., Hwang, J.Y., Kim, S.W., Kim, S.I.: Suppression of bipolar conduction via bandgap engineering for enhanced thermoelectric performance of p-type Bi0. 4Sb1. 6Te3 alloys. J. Alloys Compd. **741**, 869–874 (2018)
25. Lundstrom, M.: Notes on bipolar thermal conductivity (2017). https://nanohub.org/groups/ece656_f17/File:Notes_on_Bipolar_Thermal_Conductivity.pdf
26. Foster, S., Neophytou, N.: Effectiveness of nanoinclusions for reducing bipolar effects in thermoelectric materials. Comput. Mater. Sci. **164**, 91–98 (2019)
27. Kresse, G., Hafner, J.: Phys. Rev. B **47**, 558 (1993)
28. Kresse, G., Hafner, J.: Phys. Rev. B **49**, 14251 (1994)
29. Clark, S.J., Segall, M.D., Pickard, C.J., Hasnip, P.J., Probert, M.J., Refson, K., Payne, M.C.: Zeitschrift fuer Kristallographie **220**(5–6), 567–570 (2005)
30. Giannozzi, P., Baroni, S., Bonini, N., Calandra, M., Car, R., Cavazzoni, C., Ceresoli, D., Chiarotti, G.L., Cococcioni, M., Dabo, I., Dal Corso, A., Fabris, S., Fratesi, G., de Gironcoli, S., Gebauer, R., Gerstmann, U., Gougoussis, C., Kokalj, A., Lazzeri, M., Martin-Samos, L., Marzari, N., Mauri, F., Mazzarello, R., Paolini, S., Pasquarello, A., Paulatto, L., Sbraccia, C., Scandolo, S., Sclauzero, G., Seitsonen, A.P., Smogunov, A., Umari, P., Wentzcovitch, R.M.: J. Phys.: Condens. Matter **21**, 395502 (2009)
31. Poncé, S., Margine, E.R., Verdi, C., Giustino, F.: EPW: electron–phonon coupling, transport and superconducting properties using maximally localized Wannier functions. Comput. Phys. Commun. **209**, 116–133 (2016)
32. Neophytou, N., Kosina, H.: Effects of confinement and orientation on the thermoelectric power factor of silicon nanowires. Phys. Rev. B **83**(24), 245305 (2011)
33. Foster, S.: Ph.D. thesis (2019) University of Warwick
34. Rode, D.L., Fedders, P.A.: Electron scattering in semiconductor alloys. J. Appl. Phys. **54**(11), 6425–6431 (1983)
35. Fischetti, M.V.: Monte Carlo simulation of transport in technologically significant semiconductors of the diamond and zinc-blende structures. I. Homogeneous transport. IEEE Trans. Electron Devices **38**(3), 634–649 (1991)
36. Matthiessen, A., Vogt, C.: On the influence of temperature on the electric conducting-power of alloys. Philos. Trans. R. Soc. Lond. **154**, 167–200 (1864)
37. Neophytou, N., Wagner, M., Kosina, H., Selberherr, S.: Analysis of thermoelectric properties of scaled silicon nanowires using an atomistic tight-binding model. J. Electron. Mater. **39**(9), 1902–1908 (2010)
38. Jeong, C., Kim, R., Luisier, M., Datta, S., Lundstrom, M.: On Landauer versus Boltzmann and full band versus effective mass evaluation of thermoelectric transport coefficients. J. Appl. Phys. **107**, 023707 (2010)
39. Neophytou, N., Karamitaheri, H., Kosina, H.: Atomistic calculations of the electronic, thermal, and thermoelectric properties of ultra-thin Si layers. J. Comput. Electron. **12**(4), 611–622 (2013)
40. Martin, R.M., Martin, R.M.: Electronic Structure: Basic Theory and Practical Methods. Cambridge University Press, Cambridge (2004)
41. Jacoboni, C., Reggiani, L.: The Monte Carlo method for the solution of charge transport in semiconductors with applications to covalent materials. Rev. Mod. Phys. **55**(3), 645 (1983)

Chapter 3
Monte Carlo Method for Electronic and Phononic Transport in Nanostructured Thermoelectric Materials

3.1 General

Given a certain material, in order to understand electronic transport in its nanostructured or highly disordered forms, one needs to move beyond the simplified analytical models for the scattering rates on structure irregularities that were described in Chap. 2. For this, the geometry of the nanostructure, as well as the behaviour of the carriers upon interaction with the nanostructured features, need to be described in some detail. Such geometries naturally involve a relatively large simulation domain, larger compared to the largest of the feature sizes and large compared to the mean-free-paths of carriers. Thus, they are performed using continuum models, i.e. the domain is discretised using for example a finite difference or a finite element mesh. In principle, the Boltzmann Transport Equation (BTE) can be solved in these systems, but this is extremely difficult to do analytically. One of the 'real space' techniques that are used to understand the influence of geometry on transport properties is the Monte Carlo (MC) technique. The Monte Carlo method solves the BTE statistically by simulating a large enough number of particle trajectories under certain forces and scattering rules [1]. This technique is one of the most accurate techniques used to compute transport in electronic devices (and has been developed and described in detail for transistor devices in the past [2–4]), but it is also used for phonon transport in nanostructures. This chapter will cover the steps of implementing this method for nanostructured domains. First, the general theoretical framework and the details that dictate the electronic Monte Carlo are discussed. Afterwards the phonon Monte Carlo counterpart is discussed.

© The Author(s), under exclusive licence to Springer Nature Switzerland AG 2020
N. Neophytou, *Theory and Simulation Methods for Electronic and Phononic Transport in Thermoelectric Materials*, SpringerBriefs in Physics,
https://doi.org/10.1007/978-3-030-38681-8_3

3.2 Monte Carlo Simulation Steps

The Monte Carlo method for electron and phonon transport that is often employed for devices and nanostructured thermoelectric materials is a method that uses sampling of large number of particles to describe transport of electrons/phonons within certain spatial domains, under certain underlying rules. The Monte Carlo method considers classical ray-tracing of particles, but uses quantum mechanical electronic density of states, subbands, velocities, and scattering rates based on Fermi's Golden Rule. Thus, it is classified as 'semi-classical'. Numerically, to set up a Monte Carlo simulator for thermoelectric nanostructured materials, the following steps need to be taken:

(i) *Domain definition*: The simulation domain needs to be defined, and in the case of nanostructured TE materials a discretization in the nanometer size or even smaller is needed to appropriately describe the nanoscale features in the material. The domain discretized elements need to be assigned with the material property specifics (which matters in cases of material variations), and their conduction band edge. The band profile is usually a combination of the pristine's material band edge (including the band discontinuities in the case of domains with material variations) and the electrostatic potential that is built in the material. For example, treatment of superlattice materials will require potential barriers built from band edge variations. The driving force is also introduced in the material elements, i.e. in the case that we are interested in electronic transport, a potential difference ΔV is introduced, and a drop is inserted onto the potential of the elements from the left to the right contact. The drop can be linear in the case of uniform channels, but can also vary to capture the local variation in the resistivity of the material in the case that it is composed of different materials (usually this is obtained from self-consistent solutions with the Poisson's equation). In the case of thermoelectric materials, we are interested in low-field linear response, and thus a small potential drop is applied. It needs to be small to maintain linear response, but large enough to allow for a reasonable particle current in order to accumulate enough statistics and complete the simulation in a reasonable time frame. In general, statistical noise poses problems for Monte Carlo simulations under low fields, in which case the right/left-going fluxes are very similar, and for thermoelectric material simulations hundreds of thousands, up to a million particles need to be simulated. There are two types of MC simulations that are routinely employed, the incident flux (single particle) method, and the ensemble Monte Carlo (many particle) method. In these methods the initialization of particles differs, as well as the way current is calculated. Other things such as the treatment of scattering and the free-flight description are the same. The specifics of each method are described below.

Incident flux (or single-particle) method: In the 'incident flux' (of 'single-particle') method, a large number of particles (which can be a million electrons or phonons) are initialized at both the left/right contacts, and are injected into the channel one-by-one. We let each particle propagate through the channel from left to right one-by-one, and then all particles from right to left again one-by-one. The current is computed by determining the net flux per unit time, i.e. we keep track of the time the particles need to cross the domain from left to right, and right to left. In the electron MC case, particles which are injected from the left and are backscattered to the left, or from the right and are backscattering to the right again do not contribute to current and are not accounted for. Convergence is reached when the net flux does not change when we increase the number of particles that are injected from the contracts (up to a predefined error criterion). This method works very well for steady state simulations.

Ensemble Monte Carlo (or many particle) method: In the second method, particles are initialized in the entire channel. An example of a typical superlattice TE material is shown in Fig. 3.1 for a channel material with potential barriers. The particle motion under the applied field is traced. We call this the 'many particle' method or the 'ensemble Monte Carlo' method, and it is the most common method used in electronic devices, for example. In this case the current is computed from the flux through any cross section of the material. In this method periodic boundary conditions are used, i.e. any particle that exits the domain from the right/left sides, is re-injected from the left/right at the same energy (relatively to the band edge at the point of exit), keeping the same velocity vector as well. Convergence is achieved when enough statistics are gathered such that the flux is not altered with simulation time (again up to a predefined error criterion).

(ii) *Initialization*: The simulation begins with the initialization of the carrier ener gies and the velocity direction (electronic attributes are used from here on—phonon specifics are detailed in the subsequent section).

In the many-particle, ensemble Monte Carlo method, the energy of the electrons is selected randomly from a distribution determined according to the density-of-states (here in 3D):

$$g(E) = \frac{m^*}{\pi^2 \hbar^3} \sqrt{2m^*(E - E_C)} \tag{3.1}$$

weighted by the Fermi distribution function:

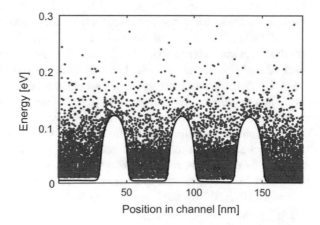

Fig. 3.1 A snapshot of electrons travelling in a channel with three potential barriers taken from a 1D ensemble MC simulation. Red dots represent right-moving electrons and blue dots left-moving electrons [5]

Fig. 3.2 The normalized
cumulative distribution
function (CDF) of the
function $g(E)f(E)$ with the
arrows indicating how a
choice of a uniform random
number in the range between
0 and 1 (y-axis) corresponds
to the carrier's energy
(x-axis) during initialization

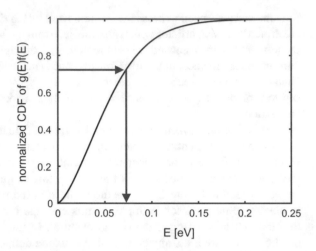

$$f(E) = \frac{1}{e^{(E-E_F)/k_B T} + 1} \tag{3.2}$$

The weighted DOS, $g(E)f(E)$, will provide a probability distribution function. This is then converted into a cumulative distribution function (CDF), which after normalization varies from 0 to 1. A uniform random number then in the range of 0 to 1 directly corresponds to the carrier's energy via the CDF. A typical CDF and the way the energy selection if performed is shown in Fig. 3.2 for $E_C = 0$ eV and $E_F = 0.05$ eV.

The selection of energy in a simple parabolic band approximation automatically restricts the velocity amplitude, and what remains now is to initialize the direction of the velocity vector. In cases were more complicated bands are used (multi-bands), then another random number will determine the band at which the carrier resides. Using correct scaling of the bands with their DOS, the idea is that a statistically accurate sampling of the carrier occupation will be constructed. To initialize the directionality of the electrons' velocity in space, we use random selection. For this, we select a random number from a uniform distribution between 0 and 2π (one angle for a 2D simulation, two angles for the 3D case—in 1D simulation the choice is only the back/forward directions). This is done for the entire number of electrons that will take part in the simulation. For example, the red/blue dots in Fig. 3.1 represent right/left moving particles. Once an electron exits the domain, using periodic boundary conditions we re-inject to from the other side, while keeping its attributes unchanged.

In the single-particle, incident flux method, we do not initialize the particle density, but the particle flux from the contacts. The reason is that in this case we do not employ periodic boundary conditions, each electron enters and exits the channel only once. Simply, we need to account for the fact that by the time a slow electron enters and exits the channel, many more faster electrons will do so. Thus, we need to weight the initializing distribution by the magnitude of the velocity of the electrons (referred to as velocity-weighted injection). Thus, the CDF is now constructed using $g(E)f(E)|v(E)|$, which is biased towards higher energies for which electrons have higher velocities. This velocity-weighted CDF, provides the new energy of the carriers that form our injection distribution. The next step in defining the injection of carriers, is to choose a direction of the carrier to be initialized. Since carriers with a larger velocity component towards the transport direction (not necessarily just velocity magnitude) will be entering the channel more frequently, the selection of the direction also needs to be weighted. For this, we use the trigonometric projection (cosine/sine function) to construct the (normalized) CDF for the angle of incidence selection.

(iii) *Free-flight*: After initialization, we allow electrons to travel in the material in 'free-flight' in the direction dictated by their velocity vector. At every discretization distance the carrier's energy and velocity is updated to the new velocity dictated by the potential drop (or raise as in the case of superlattices), i.e. the electron's kinetic energy changes as they move up/down the potential landscape. This will result in a net current flow in the presence of an electric field, as carriers will cover larger distances when their velocity increases during their flow down the potential gradient (as they will be moving higher in the bandstructure), compared to carriers moving up the potential gradient in which case their velocities will decrease. An electron's trajectory continues until it undergoes a scattering event, which ends the 'free-flight'. Scattering in TE materials can be due to phonons (acoustic and optical), ionized impurities, defects, boundaries, material variations, alloy scattering, etc.

To determine the duration of the 'free-flight', the commonly followed process is to determine the local energy dependent scattering rate. This rate is composed from the addition of all different scattering mechanisms, and the contribution of each mechanism is different at different energies. For example, the scattering rate of electrons by acoustic phonons is shown in Fig. 2.6a, and that of ionized impurity scattering in Fig. 2.6b. In the case of ionized impurity scattering the rate is different depending on the number of ionized impurities, and Fig. 2.6b shows a few examples. In the example of the channel in Fig. 3.1, in which the barriers are formed due to the spatial variation in the doping profile, the rate will change along the channel length to reflect the variation in the doping distribution.

We name the highest value, i.e. the maximum overall scattering rate as Γ_0. This will be the case at only one particular energy, as indicated in Fig. 3.3, but then we extend this value to all energies, and we assume that the total scattering rate for all energies is equal to that value, which of course is not the case. However, we label the difference from the actual rate to Γ_0 as $\Gamma_{self}(E,x)$, which is a function of energy and space. This provides the computational easiness in implementation in having an overall constant scattering rate Γ_0, and as self-scattering we consider a scattering event which does not change the electron state (momentum, energy, direction, band, etc.).

The actual 'free-flight' duration is then determined by drawing it from a Poisson distribution as:

Fig. 3.3 The total scattering rate Γ_{tot} versus energy. With the black-dashed line we also show the constant scattering rate Γ_0 and with the arrow the self-scattering rate Γ_{self} [5]

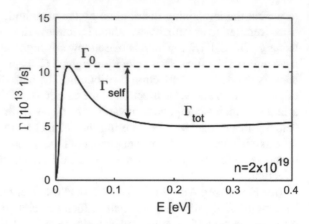

$$t = -\frac{1}{\Gamma_0} \ln(r) \qquad (3.3)$$

where r is a random number between 0 and 1, representing the interval of up to the maximum scattering rate that can be achieved. Having a constant maximum rate Γ_0 makes this drawing particularly convenient. If the variable r falls in an interval that corresponds to scattering other than self-scattering, then another random number is drawn to choose the scattering mechanism that the carrier will undergo, with the scattering rate interval separated proportionally to the scattering strength of the different mechanisms. After the scattering event then the electron's state (trajectory, energy) is updated to reflect the scattering event and the 'free-flight' is re-initiated. Acoustic phonon scattering is assumed to be elastic and isotropic, optical phonon scattering is inelastic and isotropic (consisting of optical phonon emission and absorption), and ionized impurity scattering is elastic and anisotropic. The first to be decided is the new energy of the carrier (remaining the same for elastic and changing for inelastic processes according to the energy of the optical phonon that is absorbed or emitted).

After deciding on the energy, the velocity (note that most literature uses momentum $p = mv$, rather than velocity v) of the scattered carrier is defined from the dispersion of the bands ($E = \hbar^2 k^2/2m^*$ in the parabolic case for example). To determine the scattering angle, we proceed as follows: In the case of an isotropic scattering event, the carrier scatters with equal probability in states belonging to the line/surface of the circle/sphere with radius $|v|$ in 2D/3D, where v is the velocity of the incident carrier (see Fig. 3.4). To select the scattering angle, the velocity vector in the initial direction of the particle trajectory is rotated to align with a fictitious principal x'-axis. (Once later on, the scattering angle is determined, the system is rotated back to the original basis, and the angles determined are translated into that basis). In the 2D simulation case, the polar angle (θ) is determined by randomly selecting a random number r_θ in the 0 to 1 interval from a uniform distribution and then evaluate $\theta = 2\pi r_\theta$. In the 3D case, the azimuthal angle φ is similarly randomly selected in the 0 to 2π interval using a uniform random number r_φ as $\varphi = 2\pi r_\varphi$.

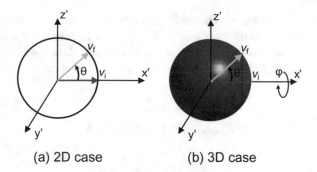

(a) 2D case   (b) 3D case

Fig. 3.4 **a** Constant energy contour/surface for a parabolic band in **a** 2D and **b** 3D and the procedure to identify the scattering angle for elastic scattering. v_i indicates the initial velocity vector and v_f the final. The initial velocity vector is rotated to align with a principal axis in order to identify the azimuthal (φ) and polar (θ) angles, and then back-rotated onto the original direction

However, the determination of the polar angle θ in 3D is more complicated. Once the azimuthal angle is determined, the polar angle cannot be uniformly chosen, as it has to reflect the fact that there are more states (area) on the sphere if one draws a circle in the $x' = 0$ plane (middle of the sphere), rather than around the dotted line as shown in Fig. 3.4b, for example. In this case, the probability to find the final velocity v_f, between a generalized polar angle θ and $\theta + d\theta$ is geometrically evaluated by:

$$P(\theta)d\theta = \frac{\sin\theta d\theta \int\limits_0^\infty \int\limits_0^{2\pi} S(k, k')d\phi k'^2 dk'}{\int\limits_0^\infty \int\limits_0^{2\pi} \int\limits_0^{\pi} [S(k, k')\sin\theta d\theta]d\phi k'^2 dk'} \tag{3.4}$$

where $S(k, k')$ is the transition rate from state k to k' as defined in 2.33. In the case of isotropic scattering $S(k, k')$ is independent of the polar angle, thus, 3.4 gives $P(\theta)d\theta = \sin(\theta)d\theta/2$. Using the CDF of this we can extract that the choice of the polar angle follows $\theta = \arccos(1 - 2r_\theta)$, where r_θ is a random number distributed between 0 and 1. The situation becomes even more complicated if the scattering is anisotropic, as the transition rate for anisotropic scattering mechanisms $S(k, k')$ has a polar angle dependence, and thus the final scattering state is not uniformly distributed on the surface of the sphere formed by the radius $|v|$ [1, 3]. In this case, 3.4 needs to be solved for the specific mechanism, and from that the appropriate CDF is formed, from which the polar angle is selected using a uniform random number. This can be performed analytically, or numerically. For example, in the case of ionized impurity scattering, small scattering angles are preferred. Scattering is then assumed to happen instantaneously, after which the electron undergoes 'free-flight' for time t, and the process is repeated until convergence (many-particle method) or until the electron leaves the simulation domain (single-particle method).

To begin the simulation, we choose the number of electrons that we will simulate, since the number of actual electrons in the material is enormously large for any simulation to be able to consider. The actual number of electrons in the volume of the material is given by the integral of the density of states over the Fermi distribution as:

$$N_{\text{device}} = V \int_{E_C}^{\infty} g(E)f(E)\,\mathrm{d}E \tag{3.5}$$

where V is the volume of the material under consideration. After we decide how many electrons we are to simulate (N_{sim}), typically a million in a domain of finite width/length, we then assign the weight for each simulated electron by defining the so-called 'super-electron' whose charge is scaled from the actual electronic charge q_0 by:

$$Q_{\text{super}} = q_0 \frac{N_{\text{device}}}{N_{\text{sim}}} \tag{3.6}$$

This number is routinely orders of magnitude larger then unity in a 3D simulation that considers 3D carrier density. However, it is computationally convenient for large scale thermoelectric systems to compute a 2D domain, and focus on detailed treatment of the nanostructured features. In this case, it is also convenient to employ 3D density of states (and still consider a small thickness in the order of a few nanometers to have a 3D volume V), but only describe the motion of electrons in the 2D plane, thus compressing the third dimension. In that case, Q_{super} can be lower than unity. In this case the third dimension is suppressed, but with enormous computational savings, allowing the execution of simulations in hours rather than days, which makes the approach useful. It is worth noting here that for strictly 1D simulations, the use of 1D DOS brings numerical difficulties is defining the Γ_0, as it turns to be very large following the van-Hove singularity in the DOS. In addition, it tends to bias the majority of electrons to be energetically placed in the singularity region, giving them very high scattering rates and very low 'free-flight' durations, making the simulation numerically very challenging. It is more convenient to use 3D DOS at all geometrical considerations, even by allowing $Q_{\text{super}} < 1$.

3.3 Nanostructuring and Boundary Scattering in Monte Carlo Simulations

During 'free-flight' in the simulation, at every discretization distance (half a nanometer or so), the velocity of the carrier is updated according to the underlying changes of the band profile. This is for example what gives electrons the forward push under the application of an electric field. In thermoelectric nanostructures, however, is also necessary at every step to check if the electron has encountered a different material, which will change its attributes, a potential barrier higher than its energy, which will force it to backscatter, a void, a boundary, etc. Once obstacles are encountered during 'free-flight', a decision needs to be taken on how the particle will behave for the rest of the 'free-flight'.

Figure 3.5 shows two examples of such scattering, first on grain boundaries, and then on the boundary of a pore (void) region. To define a randomized poly/nanocrystalline geometry one can use voronoi tessellations, an available function can be found within the MATLAB software for example, which creates grain boundaries by expanding grain areas radially outward from their initial "seeding points" until two areas meet [6], given initial input values for the number of "seeding points" and the dimensions of the domain. On the other hand, the pores/voids (or even nanoinclusions—NIs) can be done by defining the centre and radius of the NIs, and then specify the properties of the enclosed region.

Grain boundaries are in general very complicated to model accurately. They are typically associated with the introduction of the boundary resistance, which is quite complex to quantify as it involves atomistic description of materials irregularities and bond deformations, charging effects, etc. If the grain boundary or irregularity has a potential barrier associated with it (as in Fig. 3.1), then if the carrier has energy higher than the barrier it has encountered, it is allowed to flow over it. Its velocity though changes along, taking into account that the potential barrier height effectively increases the band edge. This activated over the barrier behaviour is the most common way in treating transport over grain boundaries as well. However,

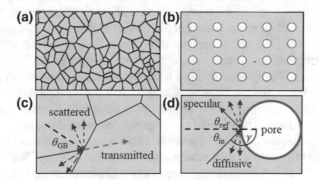

Fig. 3.5 **a** Example of a typical nanocrystalline geometry with average grain dimension $<d>$ of 75 nm. **b** Schematic of the scattering event at a grain boundary, indicating the initial angle of the particle θ_{GB} from the normal (black dashed line), grain boundaries (black lines), initial path of the particle (blue line), probable paths of the particle after scattering (red dashed lines) and probable path of the transmitted particle. **c** Example of a typical nanoporous geometry, for an ordered rectangular case, with porosity of about 10%. **d** Schematic of scattering mechanism for pore scattering, indicating the pore boundary (brown line), the initial angle of the particle θ_{in}, and potential new reflected angle of propagation θ_{ref} depending on specularity parameter, p. Probable paths of the particle after scattering for both diffusive (red dashed lines) and specular (red solid line) are also depicted

as boundary effects are in general more complex, in reality roughness exists, the local potential the electrons experience is disordered, and boundary resistance is also introduced. This is taken into account by introducing a specularity parameter p, which determines if in the case of backscattering off the boundary, the electron trajectory will be specular or diffusive. In the case of specular backscattering the angle of incidence is the same as the angle of reflection. In the diffusive case the angle of reflection is random (see schematics of Fig. 3.5).

The specularity parameter p takes values in the interval between zero and one, with zero indicating diffusive scattering, and one indicating fully specular. A random number is drawn in the interval between 0 and 1, and if what is picked is above the specified p, then a diffusive reflection is imposed for the particular carrier. More elaborate treatment also takes into account the momentum of the electron, the boundary characteristics (i.e. its roughness Δ_{rms}), and the angle of incidence of the particle with the grain boundary/pore θ_{GB} (see Fig. 3.5). A method commonly employed in the case of boundary scattering uses a relation based on diffraction theory, where in the case of no correlation length dependence, the momentum dependent specularity parameter is given by [7–9]:

$$p_{scatter} = \exp\left(-4\Delta_{rms}^2 k^2 \cos^2 \theta_{GB}\right) \tag{3.7}$$

This is commonly employed in the treatment of surface roughness scattering along the oxide/semiconductor interface in transistor devices as well. Above the scattering is more specular for smaller boundary roughness and when the perpendicular to the boundary component of the electronic wavevector, k, is small (only the perpendicular component matters here). In the case of diffusive scattering there are questions about conservation of flux along the boundary and could be considered in the choice of the angle of reflection [10], but in the case of thermoelectric materials with randomly oriented grain/grain boundaries such effects should average out, and usually are ignored [11, 12]. When the decision of what the particle will do after encountering an obstacle is taken, the attributes are changed, and the particle travels for the rest of the 'free-flight' duration in the new scattered direction. If upon scattering with the obstacle the event is considered as completely memory-loss, them a new 'free-flight' duration can be considered. However, as boundary scattering is in general very difficult to quantify, usually estimations are made for the nature of boundaries based on experimental observations. In the case of a pore, backscattering is inevitable, and in a similar manner the scattering event can be specular or diffusive. In randomized nano/polycrystalline systems, or nanoporous systems, on the other hand, as particles are randomized by geometrical means, simulations show that using a constant p, or momentum dependent p (constant Δ_{rms}), does not make much of a difference in the overall transport properties [13]. Even the considerations of fully specular, versus fully diffusive boundaries could be less important if the geometry is randomized enough and the simulation domain long enough.

After the discussions of the initializations, the 'free-flight', the scattering events, and the treatment of boundaries, we now need to describe the extraction of the current and the thermoelectric coefficients. Since this is different in each of the two methods mentioned (incident flux vs. ensemble MC), we describe each case separately.

3.4 Thermoelectric Coefficients Extraction from Monte Carlo

Current calculation in the incident flux method: In the incident flux MC method, we initialize a large and equal population of electrons in the left/right contacts, and trace them as they propagate into the material. Electrons that are injected into the channel from the left contact can propagate to the right side and exist the material, or can backscatter and exit through the left contact again. Similarly, electrons that are initialized in the right contact, can propagate through the material and exit from the left contact, or backscatter and exit from the right contact again. The left originated electrons that make it to the right side contribute to a net positive flux, whereas the right-originated ones that make it to the left side contribute to a net negative flux. The backscattered electrons do not contribute to the flux. To compute the flux we need to count the number of electrons that contribute to the fluxes, (q_{12} and q_{21} respectively), and divide each separately with the average time taken for those electrons to make it through (τ_{12}, τ_{21}, respectively). Thus, we need to be recording the time that the electrons spent in the simulation domain. The net flux is the difference between the two fluxes, and the net current is given by multiplying the flux by the assigned charge of each particle [5]:

$$I = Q_{\text{super}} \left(\frac{q_{12}}{\tau_{12}} - \frac{q_{21}}{\tau_{21}} \right) \tag{3.8}$$

In this method, each electron is being treaded independently one-by-one, and the simulation ends when all the electrons are traced through the device. One needs to check that the final current does not depend on the number of electrons, i.e. a small number of electrons will include large statistical noise. A large number is needed for convergent results, usually in the several million electrons to overcome the statistical noise associated with low field simulations. The conductivity is computed by dividing the current by the applied potential difference ΔV, which is chosen in the several mV range, small enough for linear response, and large enough to allow the gathering of significant statistics with a reasonable number of simulated carriers (for example, we find that $\Delta V = 5$ meV works fine for simulated channels of 100 nm in length).

Current calculation in the ensemble MC method: In the ensemble MC method, the electrons that participate in the simulation are all initialized randomly in the entire channel (as shown in the schematic of Fig. 3.1), and are given an initial random velocity and energy, according to carrier statistics. When the carrier reaches the end of the simulation domain, then periodic boundary conditions are applied and the carrier is re-injected from the opposite side. At steady state the flux through any given cross section of the material perpendicular to the transport direction is the same, as dictated by current continuity. The current can then be evaluated by the net flux through any cross section perpendicular to the channel multiplied by Q_{super}, and a large enough simulation time can be chosen such that enough statistics are gathered and the extracted current value remains constant over time (within a specified error criterion, usually a variation in the current of less than 1% through consecutive evaluations).

Extraction of the Seebeck coefficient: As described in Chap. 2, the Seebeck coefficient is extracted by running the simulation again under a temperature difference being the driving force. Simple examination of the equations that determine the Seebeck coefficient, however, indicate that the Seebeck coefficient is the energy weighted current flow, $S = (\langle E \rangle - E_F)/qT$, averaged over the entire channel (referred to as the energy of the current flow). Thus, there are two ways to extract the Seebeck coefficient. In practice, simulation of transport under temperature differences being the driving field can be done, but it is numerically challenging, with larger statistical error. In the case of ΔT simulations, the fluxes above and below the Fermi level have different net directionalities, which numerically complicates things when gathering reliable statistics.

On the other hand, it is more convenient to only simulate the flow when the potential difference ΔV is the driving force, but then the energy of the carriers needs to be recorded. In the case of elastic transport, in which the energy of the carriers does not change along the channel, the energy of the carriers is known from initialization, the quantity $<E> = E \times I(E)$ can be trivially extracted, and the Seebeck coefficient can be evaluated. In the more general and interesting case where optical phonons cause energy fluctuations as the carriers flow in the channel (for example in a superlattice channel as in Fig. 3.1), the situation requires more book-keeping for the carrier energies. In principle one needs: (i) for every electron that contributes to current, to keep track of their energy at all instances through their trajectory in the channel. For every electron then the average energy needs to be found at every spatial point, i.e. $E_i(x)$ in 1D for the ith electron. (ii) Then one needs to average the energy of all electrons at each spatial point, and finally (iii) average that quantity over the length of the channel. To store all information for each electron before averaging could require a significant amount of memory, but keeping only one array for the overall average energy of the current in space, and updating it by the contributions of each electron at a time, would allow significant savings.

Incorporation of tunnelling: Although semi-classical MC uses quantum mechanical bands, the ray-tracing formalism is purely classical in nature. If an electron encounters a potential barrier, no-matter how thin it is, it will undergo a full backscattering. On the other hand, we know from basic quantum mechanics, that an electron with lower energy compared to that of the barrier has a certain probability to tunnel through the barrier and make it to the other side. This probability increases as the barrier becomes thinner, and it is reasonable that it is accounted for (**yellow colormap** in Fig. 3.6). In fact, as tunnelling increases and the barrier becomes transparent,

Fig. 3.6 a Example of
quantum tunnelling through
a Gaussian-like potential
barrier (yellow colormap).
The red line denotes the
Fermi level [14]

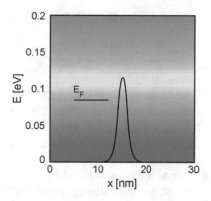

the Seebeck coefficient and the power factor both drop [14–16]. On the other hand,
we also know that electrons with energies higher compared to the potential barriers have a finite probability to backscatter. Such effects cannot be captured within
the 'mainstream' MC, but tabulated transmission probabilities for the barriers under
consideration can be extracted using quantum transport simulators such as the Non-
Equilibrium Green's Function (NEGF) method, which will be described in Chap. 4.
This can provide additional information to decide on the behaviour of the electron
when it reaches a boundary—allowing the consideration of tunnelling and quantum
reflections.

3.5   Phonon Monte Carlo Simulation Method Specifics

Because the treatment of phonons has some peculiarities compared to the treatment of
electrons, in this section we will elaborate on the numerical implementation of these
phonon Monte Carlo specific details. This formalism is also well established with
several works in the literature [11–13, 17–23], but it is useful to elaborate on some
implementation details specific to nanostructured materials. The approach specifics
described below uses Si parameters, but the method can be generally applied.

Similar to electrons, phonons are initialized according to their dispersion relation
(phonon spectrum), modified by the Bose-Einstein distribution at a given temperature. For simplicity, and since most of the heat is transferred by acoustic phonons (at
least in Si [11, 12, 20, 21, 24]), the dispersion employed in Monte Carlo simulations
(for Si) only treats the acoustic branches, which are described in a simple way by
[17]:

$$\omega(q) = v_s q + c q^2 \tag{3.9}$$

From which the phonon group velocity is extracted as:

$$v_g = \frac{d\omega}{dq} \qquad (3.10)$$

Above, v_s is the sound velocity, and three branches (or modes) are typically employed, one longitudinal, and two transverse modes. Each branch is described by its own parameters as indicated for Si in Table 3.1 and plotted in Fig. 3.7 [17, 23]. During initialization, the initial distribution function is constructed by weighting each branch with its DOS.

Phonons are assumed to be particles, alternating between 'free-flight' and scattering events. Scattering processes are due to normal phonon-phonon scattering, three-phonon scattering (also referred to as Umklapp scattering), and by interaction with geometrical features. The normal phonon-phonon and three-phonon Umklapp scattering processes are treated within the so-called single mode relaxation time approximation [12, 25–27]. In this approximation, the phonon scattering rates are given by:

$$\tau_{LA}^{-1} = B_{NU}^{LA}\omega^2 T^3 \qquad (3.11a)$$

$$\tau_{TA,N}^{-1} = B_N^{TA}\omega T^4 \qquad (3.11b)$$

Table 3.1 Parameters that are used to produce the Si acoustic phonon branches [17]

Parameter	Longitudinal acoustic (LA) branch	Transverse acoustic (TA) branch
v_s [ms^{-1}]	9.01×10^3	5.23×10^3
c [m^{2}s^{-1}]	-2×10^{-7}	-2.26×10^{-7}

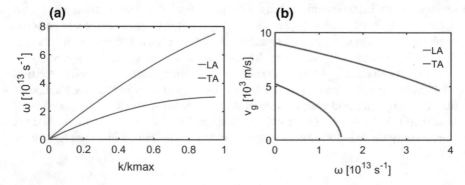

Fig. 3.7 a The fit for the dispersion relation ω and **b** the group velocity v_g obtained as in [17] for longitudinal acoustic (LA, blue lines) and transverse acoustic waves (TA, red lines). These are used for the initialization of random phonon group velocity, wave vector and other properties in Monte Carlo simulations

for normal scattering processes in the LA and TA branches respectively, and:

$$
\tau_{TA,U}^{-1} =
\begin{cases}
0 & for\ \omega < \omega_{1/2} \\
\dfrac{B_U^{TA}\omega^2}{\sinh\left(\frac{\hbar\omega}{2\pi k_B T}\right)} & for\ \omega > \omega_{1/2}
\end{cases}
\tag{3.11c}
$$

for Umklapp processes in the TA branch, where ω is the frequency, T is the temperature and $\omega_{1/2}$ is the frequency corresponding to $q = q_{max}/2$. The parameters B used in the case of Si are: $B_{NU}^{LA} = 2.0 \times 10^{-24}$ s, $B_N^{TA} = 9.3 \times 10^{-13}$ s, $B_U^{TA} = 5.5 \times 10^{-18}$ s. These equations used here are commonly used to describe relaxation time in phonon Monte Carlo simulations for Si at least [11, 12, 17, 19, 22, 23]. Using the Matthiessen's rule (which assumes that the scattering mechanisms are independent), the total relaxation time $\tau^{-1} = \tau_{LA}^{-1} + \tau_{TA,N}^{-1} + \tau_{TA,U}^{-1}$ is determined. When a phonon has spent this time τ in the simulation domain, travelling with its state's group velocity, it undergoes the next scattering event, after which the τ is reset. Notice the slight differences from the electron treatment—the alternative method described here for phonons does not define a self-scattering term and then a random time-of-flight, but defines a time of flight based on the total scattering times.

Physically, under normal scattering, the pair-system of two phonons that is involved does not change direction and the pair momentum and energy are conserved [12, 22, 28]. There is just momentum transfer between the two phonons, but this does not affect the thermal conductivity, as least directly. Indirectly, it can result in phonons with momenta that interact differently with boundaries, or undergo dissipative processes in a different manner than the original pair-system. In phonon MC simulations, however, we alter the frequency (and consequently energy and momentum) and magnitude of the velocity of the phonon, and only leave its direction unchanged (we describe the rationale for this below). Three-phonon Umklapp processes involve three phonons, typically of higher energies, and are dissipative. Two-phonons interact, and if their added momentum is larger than the length of the Brillouin zone G, then a third phonon is created, with total momentum $p_1 + p_2 - G$ and backscattered (although there are cases which is more convenient in simulations to randomize the direction after Umklapp scattering—we discuss this below). The energy lost is dissipated in the lattice, raising its temperature. This dissipative mechanism is what causes the establishment of temperature gradient in the channel material and is responsible for the thermal conductivity.

The scattering process details are as follows. If the phonon is in the TA branch its frequency determines the scattering process used. If the frequency of the phonon is high ($\omega > \omega_{1/2}$) Umklapp scattering is carried out, whereas for a low frequency phonon ($\omega < \omega_{1/2}$) a normal process is carried out. This means that for normal process, a new frequency, wavevector/momentum and velocity amplitude is chosen in the same way as is done during the initialization procedure, using the temperature of the simulation cell, but the direction remains unchanged. The relaxation time estimation [27] states that $\omega_{1/2}$ corresponds to the $G_{max}/2$ for the TA branch. For phonons in the LA branch we set that half phonons have the probability to undergo a normal scattering process, where there is no change in direction and half scatter according to Umklapp processes (we explain the rationale below) [12]. This means that we draw a random number from a uniform distribution, and check if it is below (normal) or above (Umklapp) 0.5. During the Umklapp processes we reset the phonon direction, and reset its energy using the temperature of the simulation cell where the Umklapp event happened (see below for details). During normal processes, all the participating phonons must be collinear to achieve scattering. Thus, TA mode phonon interactions would only give another TA phonon, whereas LA phonon interaction would only result in a LA phonon [11, 26].

Momentum conservation is more difficult to maintain within our simulation framework where particle-phonons travel independently. Lacroix et al. suggest an approach, by which the normal processes "approximately" preserve momentum [12]. Considering that Umklapp processes contribute to the thermal resistance directly (unlike normal processes), when the phonons scatter through an Umklapp process their direction after scattering is randomized as in the initialization procedure (rather than backscattered). Therefore, these phonons randomly scatter and contribute to the diffusion of heat. For normal processes, it is then assumed that when scattering, phonons do not change their propagation direction, only their frequency, their wave-vector and the magnitude of their velocity. By this treatment, the normal processes "approximately" preserve momentum. For a plane-parallel geometry, it seems possible to guarantee the momentum conservation in a single direction [12].

One must note that according to Holland [28] only normal processes exist for the LA branch. However, as pointed out by Lacroix et al. [12] applying this assumption means momentum has to be conserved for each scattering event involving an LA phonon, which leads to thermal conductivity values higher than the theoretical ones for temperatures between 100 and 250 K in Si. Thus, in order to ensure a more realistic momentum conservation only half of the LA phonons are scattered using normal processes and this has been validated by various works [12, 13, 22, 23, 29].

Next, we need to establish a methodology that allows for the phonon to undergo a three-phonon process and exchange energy with the lattice, even when we treat phonons as particles one at a time—which adds slightly to the complexity of the simulation scheme compared to the case of electronic Monte Carlo treatment. If a phonon undergoes a three-phonon scattering as it propagates in the channel material, its frequency and energy will be re-initialized according to the temperature of the simulation 'cell' where it scatters. The total energy of the phonons in a simulation lattice 'cell' and the temperature is connected through the relation:

$$E = \frac{V}{W} \sum_{\text{pol}} \sum_{i} \left(\frac{\hbar \omega_i}{\exp\left(\frac{\hbar \omega_i}{k_B T}\right) - 1} \right) g_p \text{DOS} \Delta \omega \tag{3.12}$$

where, ω is the frequency, T the temperature, DOS the density of states and g_p the polarization branch degeneracy. A scaling factor W is also introduced to scale the number of phonons simulated to the real population (i.e. in Si at 300 K the phonon population is ~10^5 per 10 nm^{-3} [22, 30]). Every time a phonon with energy E_{ph} undergoes a three-phonon scattering event in a 'cell' whose temperature T_{cell} corresponds to energy E_{cell}, a new phonon frequency, wave-vector and energy is re-initialized according to T_{cell}, in the same way that is done for the initialization in the contacts to begin with, i.e. according to the phonon DOS weighted by the Bose Einstein distribution function at that T_{cell}. The energy difference between this new phonon energy and the old phonon energy ΔE is dissipated in the cell. Thus, new $E_{cell}(\text{new}) = E_{cell}(\text{old}) + \Delta E$. This energy dissipation corresponds to a change in the 'cell's' temperature—the temperature can both, rise or fall, to account for the energy dissipated or extracted from the lattice. The energy difference ΔE corresponds to a temperature change ΔT as well, obtained from back-solving equation (3.12) for T. Thus, for each 'cell' in the domain we keep and update its energy E_{cell} and temperature T_{cell}.

Repeated iterations and scattering events that lead to exchange of energy with the lattice give a steady state thermal gradient between the two contacts that are placed at different temperatures T_1 and T_2. If a phonon originates from a region of higher temperature (higher energy phonon), and enters a region of lower temperature, it is more probable that after a three-phonon scattering event a phonon with a lower energy will be created (initialized) since the temperature in the new 'cell' is lower and the rest of the energy will be dissipated in the lattice, raising the lattice temperature. On the other hand, if a phonon enters a hotter region, coming from a colder region (lower energy phonon), it is more probable that after a three-phonon scattering event a phonon of higher energy will be created, absorbing energy from the lattice, and lowering its temperature. Thus, a temperature gradient is established for a continuous flow of phonons.

Scattering off geometrical irregularities such as boundaries, voids, etc., is treated in a similar manner as described for electrons earlier. The position of the phonon is checked at every discretization interval as it travels (i.e. typically every nanometer). If a boundary is encountered, then a decision has to be made how to alter the particle's trajectory. If scattering is elastic, i.e. the particle's energy remains the same, then only the direction changes. A reflection, or a transmission can be considered, depending on the nature of the boundary. In general, the scattering probability at the grain boundaries depends on the phonon wave vector, q, the roughness of the boundary, Δ_{rms}, and the angle of incidence between the phonon path and the normal to the grain boundary, θ_{GB}, as indicated in Fig. 3.5. This also determines whether an incident phonon will be transmitted to the other side or will be scattered. The probability of transmission is given by a commonly employed relation [6, 29]:

$$Tr = \exp\left(-4q^2 \Delta_{rms}^2 \sin^2\theta_{GB}\right) \tag{3.13}$$

where q in this case is the phonon wavevector and θ_{GB} is the angle between the incident phonon and the normal to the grain boundary. This means that phonons with long wavelengths (short wavevectors q) are easily transmitted through boundaries, phonons travelling perpendicular to the boundary (the larger the perpendicular component of the wavevector—the smaller θ_{GB} is) also have higher probability to be transmitted through, and smoother surfaces allow phonons to be transmitted easily as well. In the case of reflection, again depending on the nature of the boundary, the reflection can be specular or diffusive. In the case of specular reflection, the angle of reflection is the same as the angle of incidence, whereas in the diffusive boundary scattering the angle of reflection is randomized. The nature of the boundary is determined by the specularity parameter, p, with values between zero and one, with zero indicating fully diffusive boundaries and one fully specular. p can be assigned a constant value for a particular boundary, or can be chosen to be frequency dependent as well. A random number from a uniform distribution between 0 and 1 is chosen upon the incidence of a phonon on a boundary, and if that number is smaller than the specified p, then specular reflection is imposed, if larger, then diffusive reflection is imposed for that particular event.

Finally, the flux and thermal current is extracted as in the case of electronic transport, depending on what method is employed (incident flux or ensemble MC simulation methods). In the case of the incident flux, single-particle phonon MC simulation method though, two additional complications compared to electron Monte Carlo arise. The first is from the fact that the phonons change energy as they propagate through the simulation domain, thus phonons that enter from the hot side, but are backscattered and leave from the hot side again, could have dissipated/extracted energy from the domain (and vice versa for the cold contact), and these fluxes also need to be accounted for. The heat flow is flow of energy, not particle as in the electronic case. The total energy entering and leaving the simulation domain is calculated by the net sum of the corresponding phonon energies that enter/exit at the hot and cold junctions. The energy of the phonons is given by its frequency and Planck's constant i.e. $\hbar\omega_i$. Thus, the total energy crossing any junction is $E = \sum_i^{N_P} \hbar\omega_i$, where N_P is the number of phonons crossing that junction. We label the total incident energy from the hot junction as E_{in}^H and the total energy of phonons leaving the simulation domain from the hot junction as E_{out}^H. Similarly, E_{in}^C and E_{out}^C are the in-coming and out-going energies at the cold junction. We then determine the average phonon energy flux in the system as [13, 22]:

$$\Phi = \frac{\left(E_{in}^H - E_{out}^H\right) - \left(E_{in}^C - E_{out}^C\right)}{N_{sim}\langle TOF \rangle} \tag{3.14}$$

where N_{sim} is the total number of phonons simulated (usually a few million from each contact) and $\langle TOF \rangle$ is their average time-of-flight. The simulated thermal conductivity is then extracted by:

$$\kappa_s = \Phi \frac{L_X}{A_C \Delta T} \tag{3.15}$$

where A_C is the effective (scaled) cross section area of the simulation domain, which together with the scaling factor W above, are used to convert the simulated energy flux to thermal conductivity with the proper units.

The second complication arises from the presence of phonons with longer mean-free-paths (MFPs) compared to the size of the simulation domain. If we take the example of Si, phonons have mean-free-paths that span from nanometers, to millimetres, with the dominant heat carrying phonon mean-free-paths to be approximately 130–300 nm [11, 12, 20, 21, 24]. Thus, in order to be able to talk about diffusive transport for those phonons, the simulation domain should be significantly larger. On the other hand, when it comes to heat in nanostructured thermoelectric materials, the nanostructured features can be of the size of a few nanometers. Merging larger with smaller features makes computation very demanding in terms of the size of the simulation domain. In the case of the many-particle, ensemble MC simulation, where periodic boundary conditions are considered, a long MFP phonon will cross a certain cross section surface multiple times by the time a short MFP phonon crosses it, which still allows for diffusive transport for all phonons. In the incident flux, single-particle method, however, where a certain size is considered and every phonon passes through the channel only once, an extra step needs to be taken. The finite size of the simulation domain is overpassed by using the average mean-free-path λ_{ave} (135 nm in Si at 300 K, for example [24]) and the simulated channel length L_X, to scale the simulated thermal conductivity (κ_s) to the actual thermal conductivity (κ) for an infinite channel length as:

$$\kappa = \kappa_s \frac{(L_X + \lambda_{avg})}{(L_X)} \tag{3.16}$$

This scaling is important for the pristine bulk case of silicon where a large number of phonons have mean-free-paths larger than the simulation domain size [22, 24]. This is especially the case in the low temperature range where the low temperature peak of silicon is observed only after this scaling [22–24]. It allows us to simulate shorter channels, and even avoid the use of periodic boundary conditions, which simplifies the simulation considerably. In the case of nanocrystalline and nanoporous structures, on the other hand, where the scattering length is determined by the grain sizes and pore distances, this scaling is not particularly important. Note that for computational ease we can only use an average λ_{avg} value, although the mean-free-path is wavevector dependent. The separate contribution of each wavevector in the thermal conductivity in a domain where phonons undergo three-phonon processes and interchange wavevectors cannot trivially be determined. However, simulations show that adequate accuracy can be achieved. In the case of electronic transport this is not a big issue, because the electronic mean-free-paths are in general smaller compared to the domain size.

3.6 Conclusions

In this chapter we have described the implementation of electron and phonon Monte
Carlo simulations in nanostructured geometries. This approach uses ray-tracing of
particles and scattering under certain rules (some similar, some different) for electrons
and phonons. The methods were very successful in the past in providing understanding of both electron and phonon transport in materials and devices. They provide
the flexibility of describing complex geometries and nanostructured features, that
are of upmost importance in enhancing thermoelectric performance (improving the
power factor for electrons and reducing the thermal conductivity for phonons). For
both electron and phonon simulations the basic steps are: (i) to define the geometry,
(ii) initialize the particles (either in the incident flux, single-particle scheme, or in
the ensemble MC, many-particle scheme), (iii) develop a 'free-flight' algorithm that
checks at every discretization distance the properties of the underlying geometry and
whether changes/boundaries/voids, etc. are encountered, and (iv) allow for internal
scattering (electron-phonon, phonon-phonon scattering). The specific details of scattering are of course different for each particle, with the phonon simulations having to
account for heat dissipation in the lattice and the finite simulation domain. However,
one in principle uses the same basic MC principles to simulate both phonons and
electrons.

Monte Carlo simulations provide a lot of flexibility in defining the complexities
of the geometry and the rules which dictate the flow of particles. The chapter has
described the setup for single band materials. In principle, however, the technique
can include multi-band features, in which case particles are initialized in many bands,
and are allowed to scatter within many bands as well. In addition, a particle picture
is employed in MC, but we know that particles have a wave nature, and electrons
can tunnel through barriers, create resonances, interfere, etc. As we are able to select
the rules which dictate the transport details, one can employ quantum mechanical
transmission information to describe the how electrons behave once they encounter
a barrier, or wave propagation features using other software to inform Monte Carlo
on how phonons behave in the vicinity of nanoscale features. Thus, Monte Carlo
provides a powerful tool to explore the importance of nanostructuring in the performance of thermoelectric materials and could allow the identification of geometrical
features that strongly scatter the phonons, but at a much lesser degree the electrons.

References

1. Lundstrom, M.: Fundamentals of Carrier Transport. Cambridge University Press (2000)
2. Jacoboni, C., Lugli, P.: The Monte Carlo Method for Semiconductor Device Simulation, Computational Microelectronics (1989)
3. Moglestue, C.: Monte Carlo Simulation of Semiconductor Devices. Chapman and Hall (1993)
4. Laux, S.E., Fischetti, M.V., Frank, D.J.: Monte Carlo analysis of semiconductor devices: the DAMOCLES program. IBM J. Res. Dev. **34**(4), 466–494 (1990)
5. Foster, S.: Ph.D. thesis (2019) University of Warwick
6. Aksamija, Z., Knezevic, I.: Lattice thermal transport in large-area polycrystalline graphene. Phys. Rev. B **90**(3), 035419 (2014)
7. Zuverink, A.: Surface roughness scattering of electrons in bulk MOSFETS. Doctoral dissertation (2015)
8. Graebner, J.E., Reiss, M.E., Seibles, L., Hartnett, T.M., Miller, R.P., Robinson, C.J.: Phonon scattering in chemical-vapor-deposited diamond. Phys. Rev. B **50**(6), 3702 (1994)
9. Aksamija, Z., Knezevic, I.: Thermal transport in graphene nanoribbons supported on SiO_2. Phys. Rev. B **86**, 165426 (2012)
10. Soffer, S.B.: Statistical model for the size effect in electrical conduction. J. Appl. Phys. **38**(4), 1710 (1967)
11. Mazumder, S., Majumdar, A.: Monte Carlo study of phonon transport in solid thin films including dispersion and polarization. J. Heat Transf. **123**(4), 749–759 (2001)
12. Lacroix, D., Joulain, K., Lemonnier, D.: Monte Carlo transient phonon transport in silicon and germanium at nanoscales. Phys. Rev. B **72**(6), 064305 (2005)
13. Chakraborty, D., Foster, S., Neophytou, N.: Monte Carlo phonon transport simulations in hierarchically disordered silicon nanostructures. Phys. Rev. B **98**(11), 115435 (2018)
14. Neophytou, N., Kosina, H.: Optimizing thermoelectric power factor by means of a potential barrier. J. Appl. Phys. **114**(4), 044315 (2013)
15. Thesberg, M., Pourfath, M., Neophytou, N., Kosina, H.: The fragility of thermoelectric power factor in cross-plane superlattices in the presence of non-idealities: A quantum transport simulation approach. J. Electron. Mater. **45** (3), 1584 (2015)
16. Thesberg, M., Pourfath, M., Kosina, H., Neophytou, N.: The influence of non-idealities on the thermoelectric power factor of nanostructured superlattices. J. Appl. Phys. **118**, 224301 (2015)
17. Pop, E., Dutton, R.W., Goodson, K.E.: Analytic band Monte Carlo model for electron transport in Si including acoustic and optical phonon dispersion. J. Appl. Phys. **96**(9), 4998–5005 (2004)
18. Narumanchi, S.V., Murthy, J.Y., Amon, C.H.: Comparison of different phonon transport models for predicting heat conduction. J. Heat Transf. **127**, 713 (2005)
19. Pop, E., Goodson, K.E.: Thermal phenomena in nanoscale transistors. J. Electron. Packag. **128**(2), 102–108 (2006)
20. Hao, Q., Chen, G., Jeng, M.S.: Frequency-dependent Monte Carlo simulations of phonon transport in two-dimensional porous silicon with aligned pores. J. Appl. Phys. **106**(11), 114321 (2009)
21. Mittal, A., Mazumder, S.: Monte Carlo study of phonon heat conduction in silicon thin films including contributions of optical phonons. J. Heat Transf. **132**(5), 052402 (2010)
22. Wolf, S., Neophytou, N., Kosina, H.: Thermal conductivity of silicon nanomeshes: effects of porosity and roughness. J. Appl. Phys. **115**(20), 204306 (2014)
23. Wolf, S., Neophytou, N., Stanojevic, Z., Kosina, H.: Monte Carlo simulations of thermal conductivity in nanoporous Si membranes. J. Electron. Mater. **43**(10), 3870–3875 (2014)
24. Jeong, C., Datta, S., Lundstrom, M.: Monte Carlo study of phonon heat conduction in silicon thin films including contributions of optical phonons. J. Appl. Phys. **111**, 093708 (2012)
25. Klemens, P.G.: The thermal conductivity of dielectric solids at low temperatures. Proc. R. Soc. London, Ser. A **208**, 108 (1951)
26. Srivastava, G.: The Physics of Phonons. Adam Hilger, Bristol, UK (1990)
27. Han, Y.-J., Klemens, P.G.: Anharmonic thermal resistivity of dielectric crystals at low temperatures. Phys. Rev. B **48**, 6033 (1993)

28. Holland, M.G.: Analysis of lattice thermal conductivity. Phys. Rev. **132**(6), 2461 (1963)
29. Chakraborty, D., de Sousa Oliveira, L., Neophytou, N.: Enhanced phonon boundary scattering at high temperatures in hierarchically disordered nanostructures. J. Electron. Mater. **48**(4), 1909–1916 (2019)
30. Ramayya, E.B., Maurer, L.N., Davoody, A.H., Knezevic, I.: Thermoelectric properties of ultrathin silicon nanowires. Phys. Rev. B **86**(11), 115328 (2012)

Chapter 4
Non-Equilibrium Green's Function Method for Electronic Transport in Nanostructured Thermoelectric Materials

4.1 General

Nanostructuring is one of the most promising directions in reaching improved thermoelectric performance as it leads to significant reductions in thermal conductivity. The multiple and strong scattering of phonons by grains, defects, dislocations, etc., was the reason behind numerous demonstrations of ultra-low thermal conductivities, with values even below the amorphous limit. Hierarchically disordered materials, where different nature of defects are utilized to scatter phonons of mean-free-paths across the spectrum, are some of the most successful nanostructuring strategies. Of course electrons are also scattered by the same defects, but as electrons have in general significantly shorter mean-free-paths compared to phonons, they tend to suffer less. On the other hand, the various interfaces, potential barriers and defects offer opportunities for energy filtering, which could lead to power factor improvements through increases to the Seebeck coefficient. Despite the fact that the introduction of defects leads to a rather monotonic reduction in the thermal conductivity, the adverse interdependence of the electrical conductivity and Seebeck coefficient makes it harder to predict how the power factor will behave in the presence of nanostructuring. To understand this and be able to identify nanostructuring designs that are beneficial to the power factor, advanced electronic transport simulation methods are needed. The complication arises from the fact that proper treatment of electron transport needs to involve a degree of multi-scale and multi-physics component. The multi-scale aspect arises from the fact that atomic and nanoscale features need to be treated accurately within large bulk-like matrix materials. The multi-physics aspect arises from the fact that at small scales, electrons behave like quantum mechanical wave-type objects, whereas at the large scales their behaviour converges to particle-type. In addition, at the small scales around the defects a degree of ballisticity is locally encountered, whereas at the long range transport converges to classical diffusion-like.

This chapter will cover the basics of the Non-Equilibrium Green's Function (NEGF) method, which is a fully quantum mechanical electronic transport method,

N. Neophytou, *Theory and Simulation Methods for Electronic and Phononic Transport in Thermoelectric Materials*, SpringerBriefs in Physics, https://doi.org/10.1007/978-3-030-38681-8_4

and captures the essential features of transport in nanostructured materials. The usefulness of the method comes from two fronts: (i) the ability to capture and merge quantum transport to semiclassical regimes, and ballistic to diffusive regimes, and (ii) the ability to describe arbitrary geometrical features within a given channel material, thus allowing to understand how electrons interact with the nanostructuring. For the former, importantly, a numerical implementation of electron-phonon interactions is described, which allows the NEGF formalism to expand beyond the ballistic coherent regime. For the latter, the chapter will briefly provide details of geometrical features that are beneficial for power factor improvements. Since this is a well-established method (at least in the computational electronic device communities), for the purposes of this brief, I will provide the essential ingredients for a basic numerical implementation of the NEGF method in a rather 'mechanistic' way. I also only provide the basic elements of the theory and derivations of the method that have relevance to thermoelectric materials. Excellent textbooks are available, solemnly devoted on the theory behind the method, and the interested reader can refer to those, in particular the books by Datta [1–3] and relevant courses on the platform nanoHUB.org [4].

4.2   The Non-Equilibrium Green's Function (NEGF) Method

The NEGF method solves the 'open system' Schrodinger equation, and it is widely applied in the description of electron (mostly) and phonon transport in nanoscale materials and devices [1]. Over the last several years, NEGF based studies in the description of nanotransistors [5, 6], molecular devices [7], phononic effects [8], and more recently thermoelectric materials [9–11] have been performed. The strength of the method is that: (i) it accounts for quantum mechanical effects that are not captured in semi-classical methods, such as subband quantization, tunnelling, quantum reflections/transmission and their oscillations, etc., and (ii) it is rather flexible when it comes to the description of the geometrical details of the simulation domain and the underlying potential profile that exists within the material as a consequence of the nanostructuring. In terms of electronic transport in thermoelectric nanomaterials, NEGF is, thus, a very appropriate method to capture the transport details in a domain of geometrical complexity, formed by the presence of nanoinclusions, grain boundaries, superlattice potential barriers/wells, dopant variability, feature randomization, etc. The drawback is that the method is computationally expensive, scaling linearly with the channel's size, but with the third power of the simulated channel's cross section. Therefore, for nanostructured thermoelectric material considerations, its use is mostly limited to short and narrow channel materials, as well as effective mass material description level, rather than on tight-binding, or DFT material description level.

As illustrated in Fig. 4.1, the channel material, described by a Hamiltonian matrix, H, is connected to the left and right contacts, described by the self-energies $\Sigma_{1/2}$, through which electron charge will flow in/out of the device. The Hamiltonian matrix can describe the complexity of the channel material as required, i.e. including any material variations and imperfections. Figure 4.1, for example, indicates a channel on which superlattice-type barriers are introduced, or hierarchically, superlattice barriers and nanoinlcusions are introduced. In reality, such imperfections are formed by having a different material within the matrix material, or grain boundaries, or voids, etc. It is common practice in simulations, however, and much easier when it comes to coding the method on a computer, to assume a uniform Hamiltonian matrix material described by an effective mass parabolic band, and add an extra term, the potential 'U', to describe the band discontinuities between the materials. Essentially this is a local shift in the band edge E_C. Such features are important when it comes to realizing energy filtering and Seebeck coefficient improvements. In principle these different material regions can be described by a different effective mass, but we leave these complexities in the construction of the Hamiltonian matrix aside for the purposes of this brief.

While the basic implementations of NEGF consider electron transport as coherent and elastic, in which case all energies are independent of each other and electrons travel ballistically through the channel, an additional term Σ_S, called the scattering self-energy, is included to describe the interaction of electrons with the phonon

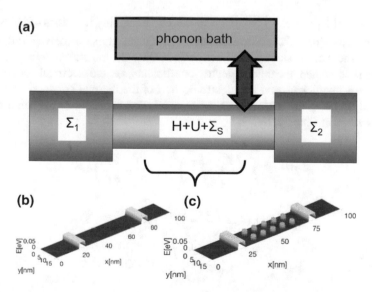

Fig. 4.1 Schematic of the system simulated by NEGF, i.e. the channel material described the Hamiltonian matrix together with an additional potential and the scattering self-energy (for example to describe electron interaction with the phonon bath). The contact self-energies describe the open boundary conditions, i.e. contacts to infinite reservoirs. Typical idealized nanostructured geometries that include superlattice and nanoinclusion barriers are shown as well

bath, i.e. electron-phonon scattering. Transport in this case becomes diffusive and inelastic in general. Thus, NEGF transits from the coherent ballistic to the incoherent diffusive regime naturally, all captured within the same formalism. This is particularly important for nanostructured thermoelectric materials, which are optimized when inhomogeneities are placed at distances similar to the electron mean-free-paths and energy relaxation distances [10, 11].

Due to the numerical complexity of electron-phonon interactions within NEGF, the majority of studies in the literature are performed to-date using ballistic, coherent conditions. Under these assumptions, electrons are injected from the two contacts as quantum mechanical wave-based objects and interfere with the channel material features coherently and elastically, subject to effects such as tunnelling, reflections, interferences, resonances, even localization. The basic NEGF equations that describe the Green's function $G(E)$, and with that the density-of-states $DOS(E)$ and the transmission function $T(E)$, for ballistic coherent transport, are:

$$G(E) = [(E + i0^+)I - H - \Sigma_1 - \Sigma_2]^{-1} \qquad (4.1a)$$

$$DOS(E) = \frac{1}{2\pi}\text{Trace}\{i[G(E) - G(E)^+]\} \qquad (4.1b)$$

$$T(E) \equiv \text{Trace}[\Gamma_2 G \Gamma_1 G^+] \qquad (4.1c)$$

with $\Gamma_{1/2} = i\left[\Sigma_{1/2} - \Sigma_{1/2}^+\right]$ being the so-called broadening functions of leads 1 and 2. The self-energies are related to elements of the Green's function between sites at the leads, and more discussions on that will follow. Once the transmission function at all energies is computed, the thermoelectric coefficients, i.e. the electrical conductivity, the Seebeck coefficient, and the electronic part of the thermal conductivity, can be evaluated by using the moments of the transmission function, in the same way as was described in Chap. 2, by:

$$I^{(j)} = \int_{-\infty}^{+\infty} \left(\frac{E - E_F}{k_B T}\right)^j T(E)\left(-\frac{\partial f}{\partial E}\right) dE \qquad (4.2a)$$

$$G = \left(\frac{2q^2}{h}\right) I^{(0)} \; [1/\Omega] \qquad (4.2b)$$

$$S = \left(-\frac{k_B}{q}\right)\frac{I^{(1)}}{I^{(0)}} \; [V/K] \qquad (4.2c)$$

$$\kappa_e = \frac{k_B^2 T}{q^2}\left[R^{(2)} - \frac{\left[R^{(1)}\right]^2}{R^{(0)}}\right] \qquad (4.2d)$$

4.3 1D Numerical Implementation of NEGF

We describe first the numerical implementation of the NEGF method in 1D, where the channel is assumed to be a simple mono-atomic chain, as this is the easiest example. Later on we describe how this can be extended to 2D. We also start with the simple ballistic coherent case (in this case we ignore the scattering self-energy Σ_S), in which the basic NEGF equations that provide the Green's function apply.

We start by constructing the elements that form the Green's function, i.e. the Hamiltonian H and the contact self-energies $\Sigma_{1/2}$. For the purposes of this brief we discretise the Hamiltonian using the effective mass approximation, which is equivalent to the single orbital orthogonal tight-binding model. To define our channel we need an effective mass m^* and the spatially dependent band edge $E_C(x)$, which includes the potential $U(x)$ terms as well. Starting from the Schrodinger's equation in 1D:

$$H\psi = E\psi \Rightarrow \left[E_c(x) - \frac{\hbar^2}{2m^*} \frac{d^2}{dx^2} \right] \psi(x) = E\psi(x) \tag{4.3}$$

we assume that the channel is a series of discretised lattice nodes with discretization a, as in the schematic of Fig. 4.2. The discretization constant is usually taken to be small compared to the feature sizes of the channel material, typically in the order of Angstroms. In that case, the second derivative of the wavefunction that appears in the Schrodinger equation can be written as the derivative of the left and right derivatives, as:

$$\left[\frac{d\psi}{dx} \right]_{x=x_n}^{\text{left}} = \frac{\psi(x_n) - \psi(x_{n-1})}{a}, \quad \left[\frac{d\psi}{dx} \right]_{x=x_n}^{\text{right}} = \frac{\psi(x_{n+1}) - \psi(x_n)}{a} \tag{4.4a}$$

$$\left[\frac{d^2\psi}{dx^2} \right]_{x=x_n} \rightarrow \frac{\frac{\psi(x_{n+1})-\psi(x_n)}{a} - \frac{\psi(x_n)-\psi(x_{n-1})}{a}}{a} = \frac{\psi(x_{n+1}) - 2\psi(x_n) + \psi(x_{n-1})}{a^2} \tag{4.4b}$$

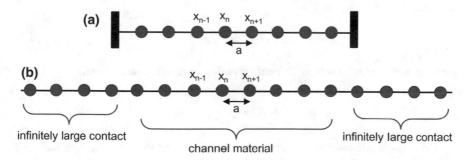

infinitely large contact

channel material

infinitely large contact

Fig. 4.2 **a** The lattice nodes that the discretised 1D Hamiltonian is based on in the orthogonal single-orbital tight-binding approximation (effective mass approximation) under closed boundary conditions. **b** The open boundary conditions system that is connected to infinitely extended reservoirs

Considering the entire discretised domain, and inserting this into the Schrodinger's equation above, we can construct its discretised matrix form. By setting $t \equiv \hbar^2 / 2ma^2$, the Schrodinger's equation in the discretized matrix form can be written as:

$$H_{1D}\psi = E\psi \Rightarrow \begin{pmatrix} E_{C1} + 2t & -t & 0 & 0 \\ -t & E_{C2} + 2t & -t & \vdots \\ 0 & -t & \ddots & -t \\ 0 & \cdots & -t & E_{Cn} + 2t \end{pmatrix} \begin{pmatrix} \psi_1 \\ \psi_2 \\ \psi_{...} \\ \psi_n \end{pmatrix} = E \begin{pmatrix} \psi_1 \\ \psi_2 \\ \psi_{...} \\ \psi_n \end{pmatrix}$$

(4.5)

where the first part is the discretized 1D Hamiltonian. The off-diagonal terms are the coupling terms, defining the 'easiness' of an electron to hop from one lattice site to the next. This is equivalent to the tight-binding approximation applied to a chain of atoms with spacing a, and one orbital per atom. In that terminology, the off-diagonal terms define the nearest neighbour hopping matrix elements $-t \equiv -\hbar^2 / 2ma^2$ and the diagonal elements are the so-called on-site energies $\varepsilon_S(x) \equiv E_C(x) + \hbar^2 / ma^2$.

This represents the 'closed system', i.e. infinitely high potential barrier walls exist at the first and last nodes, and the eigenvalues of this Hamiltonian are a close approximation to the eigenvalues of the particle-in-a-box problem. On the other hand, imposing periodic boundary conditions on the elementary unit cell of this system leads as to the (parabolic) dispersion relation.

On the other hand, the 'open-system' that NEGF describes through the proper choice of the contact self-energies $\Sigma_{1/2}$, opens the boundary conditions of the closed-system Hamiltonian by coupling it to infinitely extending reservoirs that are assumed to be in equilibrium, as schematically shown in Fig. 4.2b. The infinite wall boundary conditions of the clamped Hamiltonian are now replaced by contacts to infinitely long reservoirs, and the modification of the boundary conditions to mimic this is accounted by the self-energies. So the question that remains now is how does one construct the self-energies? For this, it is particularly easy in 1D, in which case the self-energies are simply representing outgoing waves as the appropriate boundary condition at the leads, as:

$$\Sigma_{1/2}(E) = -te^{ika},$$

(4.6)

with the wavevector is simply defined assuming parabolic bands from:

$$E = \frac{\hbar^2 k^2}{2m^*},$$

(4.7)

The contact self-energies are thought as modifications to the boundary conditions of the Hamiltonian to transform it from the 'closed' to the 'open' system, and thus, they are added only on the first and last elements of the Hamiltonian matrix, the elements that are actually attached to the contacts. The energy enters on all diagonal elements, and the Green's function in the matrix form is then:

$$
G(E)
$$

$$
= \begin{pmatrix}
(E + i0^+) - (E_{C1} + 2t + \Sigma_1) & t & 0 & 0 \\
t & (E + i0^+) - (E_{C2} + 2t) & t & \vdots \\
0 & t & \ddots & t \\
0 & \cdots & t & (E + i0^+) - (E_{Cn} + 2t + \Sigma_2)
\end{pmatrix}^{-1}
$$

$$(4.8)$$

The introduction of the infinitesimal imaginary positive number $i0^+$ on the onsites of the Hamiltonian has a very important role. When the energy E coincides with the energy band of the contacts, there are two solutions to the Green's function, corresponding to outgoing and incoming waves in the contacts, which lead to the retarded and the advanced Green's functions. The $i0^+$ corresponds to the retarded one that represents the extraction of electrons from the contact. Mathematically, it pushes the poles of G to the lower half plane in complex energy to obtain the retarded, rather than the advanced Green's function. Numerically it is a positive infinitesimal times an identity matrix of the same size as G. We choose that infinitesimal to be 10^{-10} or less for computational purposes. This term broadens the states, and could help in achieving better convergence in more general channel cases that will be discussed below (2D or 3D).

Note that the 1D NEGF implementation essentially assumes a single subband which begins at energy E_C and extends to the range of energies simulated, the $DOS(E)$ includes the well-known van Hove singularity, and the transmission is unity for energy greater than E_C and zero below (per spin channel).

4.4 Beyond 1D—Numerical Implementation of NEGF for Channels of Finite Cross Section

We now move forward to a more realistic channel consideration, namely one with extended width (the '2D' implementation). Note that as nanostructured materials do not have an invariant direction, 3D simulations are what is needed to examine them. Due to the very large computational complexities for 3D simulations, however, any such simulation will have been limited to channel materials of a few nanometers in cross section sizes. Thus, 2D is a compromise between accuracy and computational efficiency. What we describe below is essentially a channel with finite width, in addition to finite length, with the intend to have a geometrical domain which we can nanostructure—however, we still call this the '2D' channel despite not having an infinitely extending width dimension.

In a similar manner to the 1D implementation, we need to discretise the 2D Hamiltonian, and couple it to open contacts through the contact self-energies. The 2D Hamiltonian is given by:

$$H_{2D} = -\left(\frac{\hbar^2}{2m_x^*}\frac{\partial^2}{\partial x^2}\right) - \left(\frac{\hbar^2}{2m_y^*}\frac{\partial^2}{\partial y^2}\right) + U(x,y) \tag{4.8}$$

We discretise the domain in 2D by using appropriate lattice constants in x- and in y-directions as shown in the schematics of Fig. 4.3. In a similar manner as we have earlier discretised the 1D Hamiltonian, we now discretize the 2D Hamiltonian which leads to a matrix in the nearest-neighbour approximation as:

$$H_{2D} = \begin{pmatrix} h_{00} & h_{01} & 0 & 0 \\ h_{10} & h_{11} & h_{12} & \vdots \\ 0 & h_{21} & \ddots & h_{N-1,N} \\ 0 & \cdots & h_{N,N-1} & h_{NN} \end{pmatrix}_{NM \times NM} . \tag{4.9}$$

Each of the diagonal h_{ii} blocks in the 2D Hamiltonian represent the tri-diagonal y-direction columns of discretization nodes in Fig. 4.3, and the off-diagonal blocks

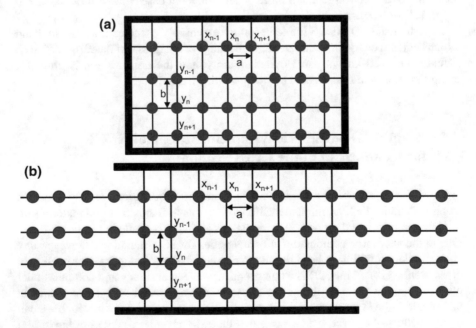

Fig. 4.3 **a** The lattice nodes that the discretised effective '2D' Hamiltonian is based on in the orthogonal single-orbital tight-binding approximation (effective mass approximation) under closed boundary conditions. **b** The open boundary conditions system that is connected to infinitely extended reservoirs

represent the couplings of those diagonal blocks to their left/right neighbours. This makes the overall matrix penta-diagonal in the nearest neighbor implementation. The size of each of the N blocks is then M × M, where M is the number of discretization nodes in the y-direction, perpendicular to transport. The entire matrix has size of (NM) × (NM). As noted, however, most elements are zero, making it a banded, block-diagonal matrix (penta-diagonal). As a banded matrix, rather than a full matrix, it makes it computation easier to solve, as numerical linear algebra solvers exist to efficiently deal with such matrices. Each diagonal block h_{ii} is written in the same way as the 1D Hamiltonian block above (but it is understood that it is now a column in the '2D' channel, rather than a 1D row channel):

$$
h_{ii} = \begin{pmatrix} E_{C1} + 4t & -t & 0 & 0 \\ -t & E_{C2} + 4t & -t & \vdots \\ 0 & -t & \ddots & -t \\ 0 & \cdots & -t & E_{Cn} + 4t \end{pmatrix}_{M \times M} \tag{4.10}
$$

The off-diagonal blocks h_{ij} (for $i \neq j$), that connect the row blocks, are simply given by:

$$
\tau \equiv h_{ij} = \begin{pmatrix} t & 0 & \ldots & 0 \\ 0 & t & \ldots & \vdots \\ 0 & \ldots & \ddots & 0 \\ 0 & \cdots & 0 & t \end{pmatrix}_{M \times M} \tag{4.11}
$$

Thus, the banded matrix has the diagonal elements, and 2 off-diagonal elements on top/bottom of the diagonal elements. Notice the difference on the number of the on-side block h_{ii} 't' elements, which is '4' rather that '2' as in the case of the purely 1D case above. This is because upon nearest-neighbour discretization in the 2D case we end up with four neighbours per side, two in the x- and two in the y-direction, rather that only two as in the 1D case.

The majorly different numerical point in the construction of the Green's function components (as in (4.1a, 4.1b, 4.1c) above), however, is the formulation of the contact self-energies $\Sigma_{1/2}$. In this case, the simple outgoing wave boundary condition expression, $\Sigma_{1/2}(E) = -te^{ika}$ used the 1D case does not work. The proper formulation for the construction of the self-energies is to 'couple' the contact to the surface Green's function, $g_s(E)$, given by:

$$
\Sigma_{L/R}(E) = h_{01}g_s(E)h_{10} = \tau_{L/R}g_s(E)\tau_{L/R}^+ \tag{4.12}
$$

where the surface Green's function is given by using the first/last elements of the 2D hamiltonian, h_{11}, h_{NN}, and the coupling matrix blocks as:

$$
g_s(E) = [(E + i0^+)I - h_{00} - \Sigma_{L/R}]^{-1} = [(E + i0^+)I - h_{00} - \tau g_s \tau^+]^{-1} \tag{4.13}
$$

Equation (4.13) above is non-linear, in the sense that g_S is a matrix block found in both the left/right hand sides of the equation. The common method to solve and isolate the surface Green's function $g_S(E)$, and then construct the self-energies using (4.12), is to solve it iteratively, routinely using the Sancho-Rubio iterative scheme [12]. Essentially, an initial value of the $g_S(E)$ is assumed, usually as an initial guess we use the inverse of the first/last on-side block elements of the Hamiltonian $g_{S,1} = h_{00}^{-1}$ ($g_{S,2} = h_{NN}^{-1}$). Afterwards (4.13) is solved iteratively, until convergence is achieved upon some criterion. The criterion used is usually a check on all elements of $\Sigma_{1/2}$ so that they change less than a small energy amount between consecutive iterations (usually a very small number like 10^{-4} eV is used). Note that the surface Greens function, and $\Sigma_{1/2}$ are full matrices, and in contrast to the isolated Hamiltonian which is Hermitian, the self-energies are complex symmetric matrices. It is clear now why the computational burden of NEGF scales with the third power of the size of the channel's width, M^3. At every iterative step an inversion of a full matrix of size M takes place. Given that 10's even 100's of inversions can be required at every energy, the computational cost is determined by these operations.

4.5 Numerical Implementation of Electron-Phonon Scattering Within NEGF

The description of the methodology up to this point included the ballistic, coherent treatment of transport within NEGF. Note that the term 'ballistic' means that electrons are transported uninterrupted through the channel, without scattering off any obstacles such as impurities, deformations, phonons, etc. This essentially means that the potential term $U(x, y)$ in the Hamiltonian is uniform (or zero), and the only self-energies included are the ones connecting the channel to the infinitely extending contacts, which are also assumed to be in thermodynamic equilibrium. The term 'coherent', means that the electrons do not undergo dephasing scattering processes, such as phonon scattering. For example, the presence of impurity scattering, within coherent transport, leads to the well described effect of Anderson localization once certain geometrical conditions are satisfied, in which case the transmission is reduced exponentially with the channel length. In this case we have coherent, but not ballistic transport. In realistic conditions, however, under room temperature, the presence of phonons will lead to phase randomization, which will lead to diffusion in which case the transmission is reduced linearly with channel length (rather than exponentially as in the case of localization). Notice that NEGF simulates the exact channel that the Hamiltonian describes, i.e. it will provide the 'conductance, G' of the specific channel, and not the 'conductivity, σ', which is the pristine material's property.

It is thus important to account for phase incoherent scattering mechanisms (electron-phonon scattering), especially in the case of thermoelectric studies for power generation, which are performed at elevated temperatures. Although the presence of impurities, material variations, etc. is treated simply by adding the corresponding potential $U(x, y)$ on the on-site Hamiltonian elements, or altering the parameters that form the Hamiltonian elements (masses, couplings, etc.), phonon scattering is more complicated. The reason is that phonons are actually not written as a part of the Hamiltonian to begin with—they are thought as the phonon bath, an external system which interacts with the electrons. To include electron-phonon interactions, we utilize the so-called 'self-consistent Born approximation'. Again, here we present the implementation of this rather that detailed theory, as that can be found in several places in the literature [1, 5, 11]. We treat phonon scattering by introducing an additional self-energy Σ_{scatt}, with the difference that now this is formed by terms that are added on all diagonal elements of the Hamiltonian (as a first order implementation this is referred to as local scattering, whereas non-local scattering will have insertions in off-diagonal elements as well):

$$G(E) = [(E + i0^+)I - H - \Sigma_1 - \Sigma_2 - \Sigma_{scatt}]^{-1} \qquad (4.14)$$

While the contact self-energies $\Sigma_{1/2}$ in (4.14) above can be left as the ones derived under the coherent formulation earlier, the scattering self-energy takes a bit of an effort to formulate and compute. Assume a state at energy E. Then the interaction with the phonon bath will lead to in-scattering processes of electrons from other states into that particular state at E, but also out-scattering of electrons that already reside at that state to flow out to other states. Either process will be accompanied with phonon emission or absorption. In a similar picture to the BTE described in previous Chapters, these in/out scattering processes of the electron interaction with the phonon bath can be described as:

$$\Sigma_{scatt}^{in}(j, j, E) = D_0(n_\omega + 1)G^n(j, j, E + \hbar\omega) + D_0 n_\omega G^n(j, j, E - \hbar\omega)$$
$$\Sigma_{scatt}^{out}(j, j, E) = D_0(n_\omega + 1)G^p(j, j, E - \hbar\omega) + D_0 n_\omega G^p(j, j, E + \hbar\omega). \qquad (4.15)$$
$$\text{phonon emission} \qquad \text{phonon absorption}$$

where the G^n and G^p stand for the electron/hole correlation functions, essentially related to the electron/hole densities, and n_ω is the number of phonons at $\hbar\omega$. The in-scattering into a state of energy E is composed of in-scattering from electrons at energy $E + \hbar\omega$ (their density described by G^n at that energy) that emit a phonon and lose energy to reach E, plus the flow from electrons at energy $E - \hbar\omega$ (their density described by G^n at that energy) that absorb a phonon and gain energy to reach E. Similarly, the out-scattering from a state of energy E is composed of electrons out-scattering from E into states at energy $E - \hbar\omega$ (their empty state density described by G^p at that energy) that emit a phonon and lose energy, plus the flow from electrons at energy E out-scattering into states of energy $E + \hbar\omega$ (their empty state density described by G^p at that energy) that absorb a phonon and gain energy. D_0 is a parameter that can be used to adjust the scattering strength (or electron-phonon coupling strength), usually to achieve a certain scattering mean-free-path (we will discuss later on how this is achieved). The values it usually takes for this are $D_0 \sim$ 0.001 eV^2, which correspond to mean-free-paths of several nanometres [10]. This is related to the deformation potential as is used in the extraction of the scattering rates in Chap. 2, but it is not directly the same—see for details [5].

In the case of electron scattering with acoustic phonons, the process is elastic, and thus, there is no change in the energy of the carrier during in-scattering, or out-scattering processes. In this case $\hbar\omega \to 0$, and both $D_0(n_\omega + 1)$ and $D_0 n_\omega \to D_A$, where D_A is now the scattering strength for acoustic processes, and the relevant in/out-scattering self-energies are now given by:

$$\Sigma_{scatt}^{in}(j, j, E) = D_A G^n(j, j, E)$$
$$\Sigma_{scatt}^{out}(j, j, E) = D_A G^p(j, j, E) \qquad (4.16)$$

where both phonon emission and absorption processes are now combined into a unified strength D_A.

We have now expressed the in/out scattering processes in terms of the electron/hole correlation functions G^n and G^p, but we still need to compute those, as well as we need to extract the Σ_{scatt} that enters the Green's function equation. We compute the electron/hole correlation functions from the Green's function and the self-energies of the entire in-flow (for G^n), and out-flow (for G^p) that the electron states are experiencing as:

$$G^n(E) = G\Sigma^{in}G^\dagger = G\left(\Sigma_1^{in} + \Sigma_2^{in} + \Sigma_{scatt}^{in}\right)G^\dagger$$
$$G^p(E) = G\Sigma^{out}G = G\left(\Sigma_1^{out} + \Sigma_2^{out} + \Sigma_{scatt}^{out}\right)G^\dagger \qquad (4.17)$$

The overall in-flows and out-flows are the summation of the in-flows from the two contacts (1/2) and the scattering self-energies. The electron-hole correlation functions depend on the overall in-flow/out-flow, which depends on the in-flow from the filled contact states and out-flow into the empty contact states, as well as the in-flow/out-flow that the electron-phonon interaction brings. To compute the filled/empty states it requires to consider the Fermi distributions of the contacts, which now need to be treated within the entire formalism. In contrast, in the coherent regime earlier, once the transmission was extracted, the current can be extracted using (4.2a) with the use of the Fermi functions a posteriori. Now, however, phonon scattering depends on the occupancy of the different states for the in/out scattering processes, and has to be treated at the same level as the rest of the quantities in the simulation. Thus, the in/out-flow self-energies that are used in (4.17) for G^n/G^p for each contact (1/2) are:

$$\Sigma_{1/2}^{in} = \Gamma_{1/2} f\left(E - E_{F(1/2)}\right) = i(\Sigma_{1/2} - \Sigma_{1/2}^\dagger) f\left(E - E_{F(1/2)}\right)$$
$$\Sigma_{1/2}^{out} = \Gamma_{1/2}\left[1 - f\left(E - E_{F(1/2)}\right)\right] = i(\Sigma_{1/2} - \Sigma_{1/2}^\dagger)\left[1 - f\left(E - E_{F(1/2)}\right)\right] \quad (4.18)$$

where above E_F denotes the Fermi level, and we have used that the broadening Γ is defined by using the contact self-energies as:

$$\Gamma = i(\Sigma - \Sigma^{\dagger}) = \Sigma^{in} + \Sigma^{out} \tag{4.19}$$

One last point here, the total scattering self-energy, that enters the Green's function (4.14) above, is formed as:

$$\Sigma^{i}_{scatt}(j, j, E) = -\frac{i}{2}\left[\Sigma^{in}_{scatt}(E) + \Sigma^{out}_{scatt}(E)\right] = -\frac{i}{2}\Gamma_{scatt}(E) \tag{4.20}$$

The expression above is the imaginary part of the total scattering self-energy, which is responsible for the scattering rate, and we commonly ignore the real part, which only brings a small shift in the eigenvalues, in general without significant consequences. Once these quantities are computed, the current can be evaluated by:

$$I_{j \to j+1} = \sum_{s} \frac{ie}{\hbar} \int_{-\infty}^{+\infty} \frac{dE}{2\pi}\left[H_{j,j+1}(E)G^{n}_{j+1,j}(E) - H_{j+1,j}(E)G^{n}_{j,j+1}(E)\right] \tag{4.21}$$

where the summation is performed over the two spin channels. It is obvious, however, from the relations presented above, that the three quantities needed to describe the system (G, $G^{n/p}$, and Σ_{scatt}) are all inter-linked. G depends on Σ_{scatt}, which depends on $G^{n/p}$, which depend on both G and Σ_{scatt}, as indicated in the schematic of Fig. 4.4a below. This complex interrelation requires a self-consistent computational loop running around these quantities. The way we implement this, is to use as initial guesses: (i) for the Green's function usually the coherent result, without any scattering, (ii) from which the $G^{n/p}$ are computed, initially without scattering, (iii) and then Σ_{scatt} is obtained—now the electron-phonon scattering terms enter the computation. Then the iterations continue until convergence, as shown in the loop of Fig. 4.4b. The usual convergence criterion is the current continuity across the channel material. The current tends not to be continuous in the initial iterations, especially in the presence of electron-optical phonon scattering, which involves carriers changing their energy as they propagate in the channel. The current however, as a flux of particles, needs

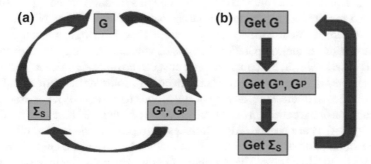

Fig. 4.4 **a** The computational self-consistent loop between the Green's function G, the electron/hole correlation functions G^{n} and G^{p}, and the scattering self-energies Σ_{scatt}. **b** Simple numerical iteration loop for implementing the self-consistent calculations

to be continuous along the length of the channel, even if the energy of the carriers differs from point-to-point along the transport direction. Thus, the current in (4.21) is computed for all nodes in the channel (j) at every iteration. The simulation is concluded when the current stops varying along the channel upon iteration progression. Usually an error criterion for the variation is set to somewhere around <1% of the current's overall amplitude.

It is worth discussing a saddle difference between the transmission extraction in the case of coherent transport, compared to incoherent transport. In the coherent regime, the energy resolved transmission function is given by (4.1c), as indicated earlier on in the Chapter, and the current through the device is calculated using the Landauer formalism as in (4.2a). The transmission is the same if it is extracted in the contacts, or any point inside the channel, as carriers travel without any alterations in their energy. The equation that provides the current in the most generic way, however, is (4.21), which is presented in the discussion of incoherent transport. In that most general transport case, we do not have a 'global' transmission explicitly, as the carriers can alter their energy as they travel along the channel (we discuss this below when we introduce transport in superlattices). In this case, it is useful to define an energy and space resolved transmission using the extracted current as [10, 11]:

$$T(E, x) = \frac{h}{2e^2} \frac{I(E, x)}{f_1(E) - f_2(E)} \tag{4.22}$$

where the term $\frac{h}{2e^2}$ is the unit of quantized transmission. We will discuss features of this generalized transmission function in the discussion of transport in superlattices below.

4.6   Recursive Green's Function Algorithm Applied in NEGF

Once all the elements that are needed for the Green's function are being constructed, (4.14) also requires an inversion of a large matrix of size NM × NM. In the 1D example where we deal with a usually smaller N × N matrix, we can use direct matrix inversion, as the computational cost is usually small. In the 2D, or even worse in the 3D case, however, the computational cost can be quite significant. Fortunately, not all elements of the inverse G matrix are needed to compute the current and maybe other quantities of interest (i.e. density of states and charge density). For example, for the calculation of the current in (4.21), only the first off-diagonal blocks of the Green's function $(j, j + 1)$ are needed. Algorithms exist to provide the needed elements by inverting small parts of the matrix, without the need to invert the entire NM × NM matrix, with O([NM]³) cost. The Recursive Green's Function iterative scheme is an efficient, and the most commonly used method for this case, which can provide the specific Green's function blocks in O(N × M³) cost, i.e. limited linearly by the size of the system, and to the third power only by the size of the perpendicular block [13]. Numerous other resources are available to explain this numerical technique in detail [14], with video resources on nanoHUB.org providing step-by-step illustrative instructions for a reader who wants to implement this [4].

4.7 Ballistic to Diffusive, and Coherent to Incoherent Transport Regimes

The NEGF formalism, as it begins with the ballistic/coherent assumptions, and incorporates electron-phonon scattering resulting to diffusive/incoherent transport, can capture the two different regimes on the same level. Depending to the electron-phonon mean-free-path for scattering, a channel of certain length can be ballistic, semi-ballistic, or diffusive. The transmission (and current) through the same structure will be different depending on the assumptions used. The transmissions for four different simulation cases are shown in Fig. 4.5 as illustration of the differences in transport properties. Here, a 2D channel is assumed, having width 30 nm and length 60 nm (left to right). The first simulation assumes ballistic/coherent transport in a uniform pristine channel. The second assumes coherent transport in a channel with embedded circular nanoinclusions. The nanoinclusions are modelled at first order as potential barriers of height $V_B = 0.1$ eV. The way this is included in the calculation is to add this barrier energy on the on-site elements of the Hamiltonian, i.e. a space varying band edge $E_C(x, y)$ is considered. This is a typical method followed in nanostructured materials to reduce their thermal conductivity through the increase of phonon-boundary scattering. The third and fourth simulations use the same two structures (pristine and with nanoinclusions), but assume electron-acoustic phonon

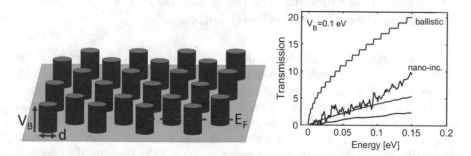

Fig. 4.5 The transmission functions for four simulation examples for a '2D' geometry: (i) A pristine material under coherent transport (blue staircase line), (ii) a material in the presence of nanoinclusions with barrier height $V_B = 0.1$ eV still under coherent transport (black non-uniform line), (iii) a pristine material under electron-phonon scattering transport conditions (blue linear line), and (iv) a material in the presence of nanoinclusions under electron-phonon scattering transport conditions (black smoothened line)

scattering as well. Here the strength of the phonon scattering is adjusted to achieve a mean-free-path of $\lambda_e = 15$ nm (we will discuss how to achieve this later on).

The transmissions in the four cases are shown in Fig. 4.5. The blue lines indicate the pristine channel simulations, whereas the black lines the simulations in the channel with the nanoinclusions. NEGF simulates the channel that is described by the Hamiltonian, and a channel with finite width will experience subband quantization, which will be captured in the ballistic transmission function as a staircase, where a step is added every time a new subband is encountered (**upper blue line**). On the other hand, in the diffusive case for the pristine channel, the transmission shape tends to approach the linear shape of the Transport Distribution Function as in the Boltzmann Transport Equation (**blue linear line**), an indication that transport in the channel is diffusive.

The coherent ballistic transmission in the structure with nanoinclusions, however (**black non-uniform line**), includes a large number of oscillations, originating from the multiple coherent reflections off the nanoinclusions. These are smeared out in the presence of electron-phonon interaction, which results in electrons losing their phase, in which case the coherence that leads to the oscillations is weakened (or eliminated), and the transmission is smoothened (**black smoothened line**). The boundary scattering and electron-phonon transport become in this case somewhat independent scattering mechanisms, which allows one to compute the overall scattering and the resultant transmission using Matthiessen's rule. In principle, however, the flexibility of NEGF allows the better treatment of channel geometries and features of sizes smaller, similar, or larger than the characteristic length scales of electron mean-free-paths (and coherence lengths), all under the same computational framework.

4.8 Simulation Procedure for the Extraction of the Thermoelectric Coefficients

As NEGF simulates the exact channel described by the Hamiltonian, it provides the current from which we extract the conductance σ, rather than conductivity G. Thus, to convert NEGF results into conductivity, we need to scale them using geometrical considerations. For example, in the 2D simulations described above, the extracted conductance needs to be scaled by the cross section area, A, and multiplied by channel length, L, to extract the conductivity as $\sigma = (L/A)G$. While the length and width of the simulation domain is known, an assumption needs to be taken for the thickness of the cross section, usually to obtain a meaningful conductivity value, something close to experimental observations. The Seebeck coefficient on the other hand, does not need to be scaled, as the conductance term appears in both its numerator and denominator, and thus, any geometrical factors cancel out. When we calculate the power factor we obtain GS^2, which can acquire the proper power factor units of W/mK^2 when the conductance is transformed into conductivity.

There are two ways to extract the thermoelectric coefficients of a particular channel in simulation 'experiments'. As in real experimental settings, one extracts the electrical conductivity by applying a small potential difference ΔV between the left/right terminal of the material. By measuring the short circuit current, the conductivity is extracted. For the Seebeck coefficient, a temperature difference ΔT is applied at the left/right contacts under zero ΔV, and from the open circuit voltage the Seebeck coefficient is extracted as $S = \Delta V_{OC}/\Delta T$. In simulations, equivalently, this can also be computed by $S = I_{ch}(\Delta V = 0) = G_S \Delta T$. In the simulation setting, strictly speaking, the temperature gradient has to also be applied along the channel, in which case a linear drop is usually assumed (when the thermal conductivity of the material is uniform throughout), and in that case the different local temperature affects the electron-phonon scattering rates. However, it is common that an average temperature is assumed in the determination of the scattering rates, and the same rates are used throughout the channel with minimal error. Thus, the calculation is run twice: once with applying ΔV to extract G, and once with applying ΔT to extract S.

The second way needs only the ΔV simulation, and saves us from running the ΔT simulation, by using the fact that the Seebeck coefficient is the average energy of the current flow as [15, 16]:

$$S = \frac{1}{qTL} \int_0^L \langle E(x) - E_F \rangle dx \qquad (4.23)$$

where q is the carrier charge ($q = -|e|$ for electrons and $q = |e|$ for holes), L is the length of the channel, and $\langle E(x) \rangle$ is the energy of the current flow along the transport direction, defined as:

$$\langle E(x) \rangle = \frac{\int_E I_{ch}(E, x) E dE}{\int_E I_{ch}(E, x) dE} \qquad (4.24)$$

where $I_{ch}(E, x)$ is the energy and position resolved current. The relation of the Seebeck coefficient to the average energy of the current flow can also be observed from BTE considerations as in Chap. 2 earlier, where the Seebeck coefficient can be written as:

$$S\big|_{x=x_0} = \frac{k_B}{q\sigma} \int_{-\infty}^{+\infty} \left(\frac{E - E_F}{k_B T} \right) \sigma(E) dE = \frac{1}{qT} \left(\frac{\int_{-\infty}^{+\infty} I_{ch}(E) E dE}{I_{ch}} - E_F \right) \qquad (4.25)$$

$I_{ch}(E)$ stands for the energy resolved current, which is the quantity we integrate in order to get the total current as $J = \int_E I_{ch}(E) dE$. Note that the current is constant along the channel at each cross section, however, its energy is not constant, i.e.

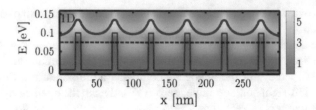

Fig. 4.6 NEGF simulation for electronic transport in a superlattice material formed by potential wells and barriers (red line). The blue line indicates the energy of the current flow, and the black line the Fermi level. The yellow/green colormap indicates high/low current energy density [17]

the charge carriers can gain or lose energy as they propagate. This happens in the presence of inelastic scattering (optical phonons) in the presence of non-uniform potential profiles in the channel, as in the usual case of superlattice thermoelectric materials shown in Fig. 4.6. Clearly, in this case, carriers absorb phonons to gain energy and overpass the barriers, and then, as the find themselves out of equilibrium, then tend to emit optical phonons and lose energy to equilibrate themselves with the well region. Thus, the energy of the current is position dependent and it is its scaled integral in the channel which provides the overall value of the Seebeck coefficient.

In [15], it was validated that the two methods of extracting the Seebeck coefficient are equivalent, which makes it easier in time consuming simulations (as the NEGF simulations) to only run the ΔV application case, and still be able to extract the Seebeck coefficient by integrating (and then scaling) the energy of the current flow over the length of the channel.

4.9 Simulation Example for Hierarchically Nanostructured Materials

The flexibility of NEGF in describing the details of the nanostructured geometry allows us to simulate complex geometries, as for example hierarchically nanostructured ones, which are promising in providing high *ZTs* through ultra-low thermal conductivities. Being able to describe electronic charge accurately, can provide reliable paths is improving the power factor as well.

Figure 4.7 depicts some idealized geometries of nanostructuring examples targeted for thermoelectric materials. A superlattice geometry is a simple first example, and one of the most promising energy filtering directions to improve the Seebeck coefficient. By filtering all low energy carriers through the use of potential barriers, the average energy of the current flow, which defines the Seebeck coefficient, increases. Power factor benefits can also be achieved by optimizing the distance between the barriers to be similar or somewhat larger compared to the energy relaxation mean-free-path of charge carriers. In this way carriers flow at high energy and the Seebeck coefficient remains high in the well. Small barrier separation (dense

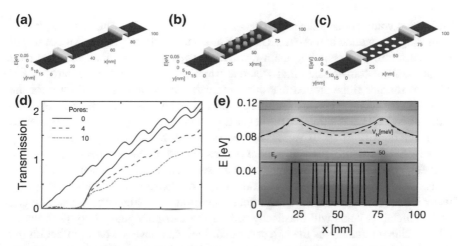

Fig. 4.7 Typical (idealized) example geometries that thermoelectric materials are nanostructured into: **a** superlattice (SL), **b** superlattice with embedded nanoinclusions NIs (hierarchical nanostructuring), **c** superlattice with embedded voids. **d** The transmission functions in a pristine material (**blue line**), in a SL material (**red line**), in a SL material with 4 NIs (**magenta line**), and in a SL material with 10 NIs (**green line**). Electron-acoustic phonon scattering conditions are assumed. **e** The current flow resolved in energy and space (**green-yellow** colormap) in a SL material with 10 NIs as in (**b**) under electron-optical phonon scattering conditions. The **blue line** denotes the Fermi level E_F. The **black-dashed line** denotes the energy of the current flow in the SL material and the **red line** the energy of the current flow in the SL material with 10 NIs. The left/right barriers denote the SL barriers, and the middle barriers the NIs [11]

barrier array), on the other hand, should be avoided as it increases the material's resistance. The use of built-in potential barriers is not only limited to superlattices, but poly/nanocrystalline materials also experience such barriers at the interfaces between grains and grain boundaries. Several works in the literature describe in detail the optimization of the well/barrier design for high power factors [11, 18–20].

The next step is the hierarchical architectures, some of which have resulted in ultralow thermal conductivities and record high ZT figures of merit. The idea behind this is that each nanostructured feature, for example atomic defects, nanoscale defects, microscale grain boundaries, targets phonons of specific mean-free-paths related to the sizes of the defects and the distance between them [21]. Such an example is shown in Fig. 4.7b, where nanoinclusions are placed in between the superlattice barriers, and in Fig. 4.7c where voids are placed in between the superlattice barriers. To include the barriers and nanoinclusions, at first order one can just raise the band edge at the specific on-site elements of the Hamiltonian to the required height, i.e. a varying $E_C(x, y)$ is used. Voids are simply modelled by raising the on-site energies to 'infinity', such that the electronic wavefunction cannot penetrate into that region. For simulation purposes, infinity is anything above several eV, and usually $V_{void} = 30$ eV will suffice.

Figure 4.7d shows the transmissions of the pristine structure (**blue line**), the superlattice structure (**red line**), and two cases of the superlattice structure with embedded voids/pores in the region between the barriers. Electron-acoustic phonon scattering is assumed in this case. The transmission functions provide a lot of useful information in understanding main features of thermoelectric transport. The shift of the onset of transmission indicates that transport begins above the barriers, but also the sharper slope around that onset indicates an advantageous feature for the Seebeck coefficient. This can be utilized if the Fermi level is placed at the level of the barrier height. This is similar to the benefits that the Seebeck coefficient experiences when the density of states varies sharply, or when the scattering times also vary sharply, but the transmission is a more generic way to express these benefits to the Seebeck coefficient in the presence of sharp features. In the presence of voids, the transmission also reveals that around the onset of transmission, voids have a very weak influence, and their degrading effect appears at much higher energies. This can be beneficial as voids will drastically reduce the thermal conductivity [22], but in a hierarchical structure as the one described, they can have a weaker effect on the electronic conductivity. The message conveyed here, however, is that a whole lot can be understood by looking at the energy resolved transmission functions, not only for the electronic conductivity, but for the Seebeck coefficient as well.

While the example of Fig. 4.7d considered only electron scattering by acoustic phonons, which is treated elastically, in reality optical phonons are present and the scattering process is inelastic. For example, Fig. 4.7e shows the current flow resolved in energy and space (green-yellow colormap) in a SL material with 10 NIs as in Fig. 4.7b under electron-optical phonon scattering conditions. The **blue line** denotes the Fermi level and the **black-dashed** line the energy of the current flow in the SL material, when only the left/right thicker SL barriers are present. The **red line** denotes the energy of the current flow in the SL material with 10 NIs (their barriers are in the middle region between the SL barriers in this cross section schematic). Clearly, electrons absorb optical phonons to make it over the SL barriers, and then tend to relax their energy to the middle potential well region by emitting optical phonons, before they encounter another barrier to pass over. In the presence of NIs this relaxation process is weakened as there is not enough energy regions for the electrons to relax into, but in general because the NIs affect carriers with energies lower that their barrier height V_N more (**red line**). This is a clear indication that the Seebeck coefficient is benefitted by the presence of the NIs—the energy of the current is higher. As non-energy relaxation is beneficial for high Seebeck coefficient, the distance between the SL barriers can be optimized such that relaxation is minimized. Of course increasing the frequency of barriers introduces barrier resistance, and an optimal distance is found for the power factor under semi-energy relaxation conditions. However, the energy of the current flow that can be extracted from NEGF, can provide useful information of how the nanostructure affects the Seebeck coefficient.

4.10 Conclusions

This chapter discussed the usefulness of the Non-Equilibrium Green's Function (NEGF) method in describing thermoelectric transport in nanostructured materials, and more specifically in hierarchically nanostructured materials. It described the implementation of the method based on the single-orbital tight-binding approximation for a channel with a finite length and width, and explained how nanoinclusions, superlattice barriers and pores can be described. It described the treatment of ballistic/coherent transport, elaborated on the numerical treatment of electron-phonon scattering and explained its importance and necessity in properly evaluating thermoelectric properties in nanostructured materials.

NEGF simulations have the advantage of being able to bridge coherent/incoherent, ballistic/diffusive, classical/quantum scales, and thus allow the treatment of electronic transport in nanomaterials in which all those effects appear and can lead to important design directions. In particular, several studies in the literature have explored this direction to optimize the thermoelectric power factor of nanomaterials, and elucidate the links between geometry and the intrinsic material properties such as mean-free-paths, electron phonon coupling strength, optical phonon energies, etc. [9, 15, 16, 19]. In particular, [18, 20] provide essential design recipes for such optimizations.

Finally, it is worth mentioning that NEGF can in a similar manner be formed to describe phonon (rather than electron) transport. In that case, instead of a Hamiltonian matrix, one constructs the Dynamical matrix, which can be built using force fields, for example. The interested reader can consult [8] for more information on this.

References

1. Datta, S.: Electronic Transport in Mesoscopic Systems. Cambridge University Press (1997)
2. Datta, S.: Quantum Transport: Atom to Transistor. Cambridge University Press (2005)
3. Datta, S.: Lessons from Nanoelectronics: A New Perspective on Transport. World Scientific Publishing Company (2012)
4. nanoHUB.org
5. Koswatta, S.O., Hasan, S., Lundstrom, M.S., Anantram, M.P., Nikonov, D.E.: Nonequilibrium green's function treatment of phonon scattering in carbon-nanotube transistors. IEEE Trans. Electron Devices **54**(9), 2339–2351 (2007)
6. Liang, G., Neophytou, N., Lundstrom, M.S., Nikonov, D.E.: Ballistic graphene nanoribbon metal-oxide-semiconductor field-effect transistors: a full real-space quantum transport simulation. J. Appl. Phys. **102**(5), 054307 (2007)
7. Pomorski, P., Odbadrakh, K., Sagui, C., Roland, C.: Nonequilibrium Green's function modeling of the quantum transport of molecular electronic devices. Theor. Comput. Chem. **17**, 187–204 (2007)
8. Karamitaheri, H., Pourfath, M., Kosina, H., Neophytou, N.: Low-dimensional phonon transport effects in ultranarrow disordered graphene nanoribbons. Phys. Rev. B **91**(16), 165410 (2015)
9. Thesberg, M., Pourfath, M., Kosina, H., Neophytou, N.: The influence of non-idealities on the thermoelectric power factor of nanostructured superlattices. J. Appl. Phys. **118**(22), 224301 (2015)

10. Foster, S., Thesberg, M., Neophytou, N.: Thermoelectric power factor of nanocomposite materials from two-dimensional quantum transport simulations. Phys. Rev. B **96**(19), 195425 (2017)
11. Vargiamidis, V., Neophytou, N.: Hierarchical nanostructuring approaches for thermoelectric materials with high power factors. Phys. Rev. B **99**(4), 045405 (2019)
12. Sancho, M.L., Sancho, J.L., Rubio, J.: Quick iterative scheme for the calculation of transfer matrices: application to Mo (100). J. Phys. F Met. Phys. **14**(5), 1205 (1984)
13. Lake, R., Klimeck, G., Bowen R.C., Jovanovic, D.: Single and multiband modeling of quantum electron transport through layered semiconductor devices. J. Appl. Phys. **81**, 7845 (1997)
14. Anantram, M.P., Lundstrom, M., Nikonov, D.: Modeling of nanoscale devices. Proc. IEEE **96**, 1511 (2008)
15. Kim, R., Lundstrom, M.S.: Computational study of the seebeck coefficient of one-dimensional composite nano-structures. J. Appl. Phys. **110**, 034511 (2011)
16. Vargiamidis, V., Thesberg, M., Neophytou, N.: Theoretical model for the Seebeck coefficient in superlattice materials with energy relaxation. J. Appl. Phys. **126**(5), 055105 (2019)
17. Thesberg, M., Kosina, H., Neophytou, N.: On the effectiveness of the thermoelectric energy filtering mechanism in low-dimensional superlattices and nano-composites'. J. Appl. Phys. **120**, 234302 (2016)
18. Neophytou, N., Zianni, X., Kosina, H., Frabboni, S., Lorenzi, B., Narducci, D.: Simultaneous increase in electrical conductivity and seebeck coefficient in highly boron-doped nanocrystalline Si. Nanotechnology **24**(20), 205402 (2013)
19. Kim, R., Lundstrom, M.S.: Computational study of energy filtering effects in one-dimensional composite nano-structures. J. Appl. Phys. **111**, 024508 (2012)
20. Neophytou, N., Foster, S., Vargiamidis, V., Narducci, D.: Nanostructured potential well/barrier engineering for realizing unprecedentedly large thermoelectric power factors. Mater. Today Phys. **11**, 100159 (2019). https://doi.org/10.1016/j.mtphys.2019.100159
21. Biswas, K., He, J., Blum, I.D., Wu, C.-I., Hogan, T.P., Seidman, D.N., Dravid, V.P., Kanatzidis, M.G.: High-performance bulk thermoelectrics with all-scale hierarchical architectures. Nature **489**, 414–418 (2012)
22. de Sousa Oliveira, L., Neophytou, N.: Large-scale molecular dynamics investigation of geometrical features in nanoporous Si. Phys. Rev. B **100**(3), 035409 (2019)

Chapter 5
Summary and Concluding Remarks

Recent developments in thermoelectricity related to the realization of ultra-low thermal conductivity nanostructured materials and the identification of a large variety of potentially high performance materials and their alloys, have led to large improvements in *ZT* across materials and temperatures. Theory and simulation are key tools to optimize materials, but also identify new directions for performance improvement. It is not surprising that after decades of inactivity in thermoelectric research, the resurgence began after a theoretical paper by Hicks and Dresselhaus, which indicated that low-dimensional materials could provide large Seebeck coefficients [1]. Despite the fact that this prediction was not to-date successfully demonstrated, it provided the spark that started the nanostructuring 'revolution' in thermoelectrics. In addition, the progress in solid state chemistry allowed the realization of myriads of materials and alloys that provide inherent low thermal conductivities, and/or bandstructures with multiple bands, valleys, degeneracies, and thus high power factors. This allowed for a large design space to be explored in the search of new thermoelectric materials.

The seemingly infinite space of possible material exploration is recently started to be scanned using machine learning techniques for high throughput calculations using DFT electronic structure data and the Boltzmann Transport Equation (BTE) theory [2, 3]. For speeding up the calculations, several attempts to identify relevant and accurate descriptors have emerged as well [3, 4]. Accurate theory and simulation is essential towards that direction. Due to the numerical complexity of identifying the relaxation times that are used within the BTE, the simplest way is the assumption of the 'constant relaxation time—CRTA' approximation. Recent sophisticated software (EPW) computes the electron-phonon relaxation times from first principles [5], but this is computationally extremely expensive to be used in high throughput calculations. On the other hand, electron-ionized impurity scattering, electron-alloy scattering, and electron-boundary scattering could be as important, if not more, compared to electron-phonon scattering.

In this brief, Chap. 2 introduced the basic features of the BTE, and how thermoelectric coefficients are derived. Then, I provided the computational details of a fully numerical scheme that extracts scattering rates using deformation potential

© The Author(s), under exclusive licence to Springer Nature Switzerland AG 2020
N. Neophytou, *Theory and Simulation Methods for Electronic and Phononic Transport in Thermoelectric Materials*, SpringerBriefs in Physics,
https://doi.org/10.1007/978-3-030-38681-8_5

theory, taking account of the energy/momentum/band dependence of all states in the Brillouin zone. This method can be thought as an intermediate way (in terms of accuracy) between the simplified 'CRTA' and the computationally expensive EPW method. It still uses deformation potentials as input parameters, but it can incorporate in a straight forward way scattering from acoustic phonons, optical phonons, ionized impurity scattering (including screening, which relies on the dielectric constant values), alloy scattering, boundary scattering, etc., in a flexible way. This middle path described is computationally more expensive than the CRTA, but much less expensive compared to the EPW method. Quantitatively, it could provide some advantages compared to the CRTA, subject to the accuracy of the deformation potentials employed. Qualitatively, however, including the correct energy/momentum/band dependence of the relaxation times, in addition to having the flexibility of allowing intra- versus inter-valley scattering, provides better understanding of transport properties, and allows reaching better design directions. Compared to first principle calculations for the scattering rates the method described in Chap. 2 could lack quantitatively, however, it has the advantage of allowing the flexibility to introduce other important scattering mechanisms. Thus, in the search for thermoelectric materials, the different transport methods compete between computational efficiency, accuracy, and flexibility, but all can prove useful in their own way. One has to understand the benefits and limitations of each one. Importantly, one reaches different conclusions regarding the most appropriate descriptors that can be used within high throughput calculations and machine learning techniques. For example, under the CRTA the large density-of-states effective mass and number of bands benefits the power factor, but light effective masses are beneficial when the energy dependence in the scattering times is included. Different ranking in materials screening techniques will be reached, so such studies need to consider such issues.

The next two Chapters focused on simulation techniques related to the second important direction in thermoelectric materials research, i.e. modelling transport in nanostructured materials. In a similar way to the extraction of the thermoelectric coefficients for complex bandstructure materials, the extraction of thermoelectric coefficients in nanostructured materials has its own challenges that cannot be all captured in one simulation technique. Ballistic to diffusive, coherent to incoherent, wave to particle regimes can appear within regions of the material under complex, hierarchical nanostructuring, and all need to be captured. In addition, length scales from atomistic, to the nanoscale, and up to the macroscale can be important, which requires atomistic treatment in some regions of the material, but continuum in others, in order to handle the computational complexities. Thus, multi-scale and multi-physics techniques are essential. Again, however, depending on the problem in hand, the choice of the different method can be more appropriate to handle the situation in the best way possible.

Chapter 3 described details for the implementation of the Monte Carlo method, which using ray-tracing can describe the motion of particles under Newton's laws and under scattering, essentially solving the BTE in a statistical manner. This can be applied to both electrons and phonons. The method is most appropriate in regimes that the carriers can be described as particles, and is suitable for treating large structures with embedded nanoscale features. The computational expense is determined linearly by the length of the simulation domain and the amount of nanostructuring. Two different flavours of this method are described, the ensemble particle method (many particle), and the incident flux method (single particle). The method is purely semiclassical, however, some quantum mechanical effects such as tunnelling though potential barriers can be easily included/approximated. In this case, pre-calculated transmission functions using quantum mechanical methods such as NEGF can be used to essentially provide a statistical probability for the electron to pass through the barrier rather than reflect. The method, however, is very flexible in including all sorts of relevant scattering mechanisms (ionized impurities, phonons, alloy scattering, etc.). The geometrical features that scatter electrons can be treated in 'real space', i.e. the geometry is constructed to reflect the actual grains, barriers, nanoinclusions, etc., in which regions the properties of the material change accordingly. Transport in nanocrystalline materials, superlattices, modulation doping materials, material variability, all such geometries can be modelled using this method, and optimization directions for the power factor can be demonstrated.

On the other side of the spectrum, Chap. 4 described the NEGF method, a purely quantum mechanical method, which begins with the purely wave, coherent nature of the electron. By including interactions with the phonon bath (electron-phonon scattering), diffusion is eventually reached. Thus, the method is capturing the multiphysics aspect of transport, and as the Hamiltonian of the channel material can be written to describe nanostructured features, it makes it ideal for thermoelectric studies. On the other hand, the method is computationally expensive, and scales with the third power of the cross section of the simulation domain. Although scattering can be included through the proper choice of self-energies, computation and convergence becomes challenging when scattering is included, and when the nanostructured geometry becomes more and more complicated. Thus, depending on the problem at hand, and whether the need to capture multi-physics issues or the need to model larger domains and more flexible scattering treatment is important, each method (Monte Carlo or NEGF) can have its advantages and disadvantages.

To sum up, the emergence of complex bandstructure materials and nanostructuring has created the need for advanced simulators that capture properly the relevant effects. The complexity of the task is tremendous, as computationally anything that can provide useful and quantitatively accurate results is expensive, and merges multiphysics, multi-scale phenomena. All these are not captured in a unified simulator, and no single method to-date can capture all important effects. Thus, multiple methods, each with its own advantages and disadvantages are utilized, and it is the job of the user to understand the strength and limitations of each method to provide relevant material/nanostructure evaluation or new design directions. The complexity of the topic is such that an essential and integrated part of the investigations is to turn back to experiments and validate simulator outcomes qualitatively and quantitatively with measured data.

This brief attempts to provide essential details for the implementation of certain computational tools that can be used in assessing the performance of the next generation thermoelectric materials. It is, however, by no means complete. A large number of details are omitted, as well as most of the theoretical background behind the methods. As materials become more and more complex, and as experimentalists are able to grasp more and more control over the synthesis of materials and their nanostructures, more physical phenomena might need to be included, and many approximations might need to be relaxed. What this brief provides is an initial guidance for the interested researcher, which will have to be gradually expanded and enrichened.

References

1. Hicks, L.D., Dresselhaus, M.S.: Phys. Rev. B **47**(24), 16631 (1993)
2. Oliynyk, A.O., Antono, E., Sparks, T.D., Ghadbeigi, L., Gaultois, M.W., Meredig, B., Mar, A.: Highthroughput machine-learning-driven synthesis of full-heusler compounds. Chem. Mater. **28**(20), 7324–7331 (2016)
3. Xing, G., Sun, J., Li, Y., Fan, X., Zheng, W., Singh, D.J.: Electronic fitness function for screening semiconductors as thermoelectric materials. Phys. Rev. Mater. **1**(6), 065405 (2017)
4. Chen, X., Parker, D., Singh, D.J.: Importance of non-parabolic band effects in the thermoelectric properties of semiconductors. Sci. Rep. **3**, 3168 (2013)
5. Poncé, S., Margine, E.R., Verdi, C., Giustino, F.: Comput. Phys. Commun. **209**, 116–133 (2016)

Index

A

Acoustic deformation potential scattering, 33

Average energy of the current flow, 75, 76

B

Ballistic transport, 27, 68

Bandstructure, 1, 2, 5, 9–11, 15, 21–23, 27–31, 33, 81–83

Bipolar, 20, 21

Boltzmann Transport Equation (BTE), 2, 3, 9–12, 15, 16, 19, 27, 33, 37, 69, 74, 75, 81, 82

Boundary scattering, 9, 23, 25, 44, 46, 54, 73, 74, 81, 82

C

Closed system, 64

Cumulative distribution function, 40

D

Density Functional Theory (DFT), 2, 3, 10, 21, 28, 60, 81

Density of states, 15, 21, 24, 25, 28–30, 38, 44, 53, 72, 78

Diffusive scattering, 46

Diffusive transport, 55

Discretization, 27, 31, 38, 41, 44, 53, 56, 63, 66, 67

Distribution function, 9–14, 16, 23, 39, 40, 50, 53

E

Effective mass, 11, 15, 33, 60, 61, 63, 66, 82

Electrical conductivity, 1, 4, 5, 59, 62, 75

Energy filtering, 1, 4, 26, 59, 61, 76

Ensemble Monte Carlo, 38, 39

F

Flux, 13, 14, 16, 17, 38–40, 46–48, 54–56, 71, 83

G

Green's function, 62, 63, 65, 67, 68, 70–73

H

Hamiltonian matrix, 61, 64, 79

Hierarchical nanostructuring, 2, 77, 82

I

Incident flux method Monte Carlo, 38, 47

Ionized impurity scattering, 9, 21–24, 32, 41–43, 81, 82

M

Many particle Monte Carlo, 38, 39

Mean-Free-Path (MFP), 55

Molecular Dynamics (MD), 2–4, 19

Monte Carlo, 4, 5, 19, 37, 38, 44, 47, 49–52, 54, 56, 82, 83

Multi-physics, 59, 82, 83

Multi-scale, 59, 82, 83

N
Nanoinclusions, 21, 25, 26, 45, 60, 61, 73, 74, 77, 79, 83
Nanostructure, 3, 19, 37, 44, 65, 78, 83, 84
Non-Equilibrium Green's Function (NEGF), 4, 5, 49, 59–65, 68, 72–74, 76, 78, 79, 83

O
Open system, 60
Optical deformation potential scattering, 23

P
Phonon bath, 61, 62, 69, 83
Phonon scattering, 2–5, 9, 15, 20–25, 33, 42, 50, 52, 53, 56, 62, 68–71, 73–75, 77–79, 81, 83
Potential barriers, 21, 25, 26, 38, 39, 45, 48, 49, 59, 60, 64, 73, 76, 77, 83
Power factor, 1, 3–5, 9, 19–21, 33, 49, 56, 59, 60, 74, 76–79, 81–83

Q
Quantum transport, 49, 60

R
Random number selection, 42

R
Raytracing, 5
Recursive Green's function, 72
Relaxation scattering time, 10, 33

S
Sancho-Rubio algorithm, 68
Screening length, 24
Seebeck coefficient, 1, 4, 5, 16, 17, 19–21, 26, 27, 48, 49, 59, 61, 62, 74–78, 81
Self-consistency, 38, 69, 71
Self-energy, 61–64, 66–71, 83
Self-scattering, 41, 42, 51
Semiclassical transport, 9
Single particle Monte Carlo, 38
Specular scattering, 25, 45, 46

T
Thermal conductivity, 1, 3–5, 18–21, 27, 51, 52, 54–56, 59, 62, 73, 75–78, 81
Time-of-flight, 51, 54
Transport distribution function, 15–17, 19, 74
Triangulation, 29–31
Tunnelling, 26, 48, 49, 60, 62, 83

U
Uniform distribution, 40, 42, 52, 54

Printed in the United States
By Bookmasters